Student Solutions Manual

for

Bettelheim, Brown, and March's

Introduction to General, Organic, and Biochemistry

Seventh Edition

Student Solutions Manual

for

Bettelheim, Brown, and March's

Introduction to General, Organic, and Biochemistry

Seventh Edition

Mark Erickson
Hartwick College

William Brown
Beloit College

Rodney Boyer
Hope College

THOMSON

BROOKS/COLE

Australia • Canada • Mexico • Singapore • Spain • United Kingdom • United States

Printed in the United States of America
2 3 4 5 6 7   07 06 05 04 03

Printer: Globus Printing

ISBN: 0-534-40179-1

For more information about our products,
contact us at:
Thomson Learning Academic Resource Center
1-800-423-0563

For permission to use material from this text,
contact us by:
Phone: 1-800-730-2214
Fax: 1-800-731-2215
Web: http://www.thomsonrights.com

Asia
Thomson Learning
5 Shenton Way #01-01
UIC Building
Singapore 068808

Australia/New Zealand
Thomson Learning
102 Dodds Street
Southbank, Victoria 3006
Australia

Canada
Nelson
1120 Birchmount Road
Toronto, Ontario M1K 5G4
Canada

Europe/Middle East/South Africa
Thomson Learning
High Holborn House
50/51 Bedford Row
London WC1R 4LR
United Kingdom

Latin America
Thomson Learning
Seneca, 53
Colonia Polanco
11560 Mexico D.F.
Mexico

Spain/Portugal
Paraninfo
Calle/Magallanes, 25
28015 Madrid, Spain

Student Solutions Manual
For
Introduction to General, Organic, and Biochemistry,
Seventh Edition

Table of Contents

Introduction to Organic and Biochemistry, Fifth Edition

Table of Contents for Organic and Biochemistry Users

Chapter 1 Matter, Energy, and Measurement

1.1 (a) $°F = \dfrac{9}{5} °C + 32 = \dfrac{9}{5} 64.0°C + 32 = 147°F$

(b) $°C = \dfrac{5}{9} (°F - 32) = \dfrac{5}{9} (47°F - 32) = 8.3°C$

1.3 $8.55 \text{ mi} \left(\dfrac{1.609 \text{ km}}{1 \text{ mi}} \right) = 13.8 \text{ km}$

1.5 $\text{mass of Ti} = 17.3 \text{ mL} \left(\dfrac{4.54 \text{ g Ti}}{1 \text{ mL}} \right) = 78.5 \text{ g Ti}$

1.7 The specific gravity of a substance is the density of this substance divided by the density of water, which is 1.00 g/mL.

$$\text{specific gravity} = 1.016 = \dfrac{d}{1.00 \text{ g/mL}}$$

$d = 1.016 \text{ g/mL}$

1.9 $(T_2 - T_1)°C = \dfrac{\text{heat}}{SH \times m} = \dfrac{230 \text{ cal}}{(0.11 \text{ cal} / g \cdot °C)(100 \text{ g})} = 21°C$

$(T_2 - T_1)°C = 21°C$

$T_2 = 21°C + 25°C = 46°C$

1.11 A major reason for the increase of the average life expectancy of humans in the last 80 years has been great progresses in medical science. Diseases that were once fatal have either been eradicated or cures developed. The causes and cures for many major diseases are now better understood and treatments more effective.

1.13 Metals have low specific heats and organic compounds have higher specific heats.

<u>1.15</u> (a) Chemical change: burning gasoline is converted to carbon dioxide and water.
   (b) Physical change: ice forming from liquid water is still H_2O, just a different state of matter.
   (c) Physical change: boiling oil is still oil, just a different state of matter.
   (d) Physical change: melting lead remains lead, just a different state of matter.
   (e) Chemical change: elemental Fe has been converted to rust, Fe_2O_3.
   (f) Chemical change: nitrogen and hydrogen converted ammonia, NH_3, involved a change in chemical composition.
   (g) Chemical change: the chemical components in food are converted to energy, carbon dioxide, and water thus changing chemical composition.

<u>1.17</u> (a) 403,000   (b) 3,200   (c) 0.0000713   (d) 0.000000000555

<u>1.19</u> (a) 2.00×10^{18} (b) 1.37×10^5 (c) 2×10^{-10} (d) 8.8×10^1 (e) 1.56×10^{-8}

<u>1.21</u> (a) 7.74×10^{-3} (b) 8.808×10^{-2} (c) 1.3022×10^2

<u>1.23</u> 3.25×10^{-5}

<u>1.25</u> (a) 3   (b) 2   (c) 1   (d) 4   (e) 5

<u>1.27</u> (a) 25,000   (b) 4.1   (c) 15.5

<u>1.29</u> (a) 10963.1   (b) 244   (c) 172.34

<u>1.31</u> (a) 1 kg = 1000 g   (b) 1 mg = 0.001 g

<u>1.33</u> (a) 100 cm   (b) 230 mL   (c) 75 kg   (d) 15 mL
      (e) 50 mg   (f) 100 mm   (g) 40 g

<u>1.35</u> Temperature conversions: $^\circ C = \dfrac{5}{9} (^\circ F - 32)$ and $K = 273 + ^\circ C$

   (a) $\dfrac{5}{9} (320^\circ F - 32) = \underline{160^\circ C}$ and $273 + 160^\circ C = \underline{433\ K}$

   (b) $\dfrac{5}{9} (212^\circ F - 32) = \underline{100^\circ C}$ and $273 + 100^\circ C = \underline{373\ K}$

   (c) $\dfrac{5}{9} (0^\circ F - 32) = \underline{-18^\circ C}$ and $273 + (-18)^\circ C = \underline{255\ K}$

   (d) $\dfrac{5}{9} (-250^\circ F - 32) = \underline{-157^\circ C}$ and $273 + (-157)^\circ C = \underline{116\ K}$

1.37 Unit conversions:

(a) $42.6 \ \cancel{kg} \left(\dfrac{2.205 \ lb}{1 \ \cancel{kg}} \right) = 93.9 \ lb$

(b) $1.62 \ \cancel{lb} \left(\dfrac{453.6 \ g}{1 \ \cancel{lb}} \right) = 735 \ g$

(c) $34 \ \cancel{in} \left(\dfrac{2.54 \ cm}{1 \ \cancel{in}} \right) = 86 \ cm$

(d) $37.2 \ \cancel{km} \left(\dfrac{1 \ mi}{1.609 \ \cancel{km}} \right) = 23.1 \ mi$

(e) $2.73 \ \cancel{gal} \left(\dfrac{3.785 \ L}{1 \ \cancel{gal}} \right) = 10.3 \ L$

(f) $62 \ \cancel{g} \left(\dfrac{1 \ oz}{28.35 \ \cancel{g}} \right) = 2.2 \ oz$

(g) $33.61 \ \cancel{qt} \left(\dfrac{1 \ L}{1.057 \ \cancel{qt}} \right) = 31.80 \ L$

(h) $43.7 \ \cancel{L} \left(\dfrac{1 \ gal}{3.785 \ \cancel{L}} \right) = 11.5 \ gal$

(i) $1.1 \ \cancel{mi} \left(\dfrac{1.609 \ km}{1 \ \cancel{mi}} \right) = 1.8 \ km$

(j) $34.9 \ \cancel{mL} \left(\dfrac{1 \ fl \ oz}{29.57 \ \cancel{mL}} \right) = 1.18 \ fl \ oz$

1.39 Yes, you would reach Ottawa within an hour

$$\text{speed} = \frac{75 \ \cancel{mi}}{1 \ hr} \left(\frac{1.609 \ km}{1 \ \cancel{mi}} \right) = 120 \ kph$$

$$\text{time to reach Ottawa} = 80 \ km \left(\frac{1 \ hr}{120 \ km} \right) = 0.67 \ hr$$

1.41 $\text{car efficiency} = \left(\dfrac{25.00 \ \cancel{mi}}{\cancel{gal}} \right) \left(\dfrac{1.609 \ km}{1 \ \cancel{mi}} \right) \left(\dfrac{1 \ \cancel{gal}}{3.785 \ L} \right) = 10.63 \ km/L$

1.43 Manganese (d = 7.21 g/mL) is more dense than the liquid (d = 2.15 g/mL) therefore it will sink. Sodium acetate (d = 1.528 g/mL) is less dense than the liquid, therefore it will float on the liquid. Calcium chloride (d = 2.15 g/mL) has a density equal to that of the liquid, therefore it will stay in the middle of the liquid.

1.45 $d_{\text{urine sample}} = m/V = \left(\dfrac{342.6 \ g}{335.0 \ \cancel{cc}} \right) \left(\dfrac{1 \ \cancel{cc}}{1 \ mL} \right) = 1.023 \ g/mL$

1.47 Water (d = 1.0 g/cc) will be the top layer in the mixture because its density is lower than dichloromethane (d = 1.33 g/cc).

1.49 Water reaches its maximum density at 4°C; therefore, by warming the water from 2°C to 4°C, the lighter crystals will float on the water with an increased density at 4°C.

1.51 While driving your car, the car's <u>kinetic energy</u> (energy of motion) is converted by the alternator to electrical energy, which charges the battery, storing <u>potential energy</u>.

1.53 $SH_{unknown} = \dfrac{heat}{m \times (T_2 - T_1)}$

$SH_{unknown} = \dfrac{2750\ cal}{168\ g\ (74° - 26°)°C} = 0.34\ cal/g \cdot °C$

1.55 $drug\ dose_{135lb} = 135\ \cancel{lb\text{-}man} \left(\dfrac{445\ mg\ drug}{180\ \cancel{lb\text{-}man}} \right) = 334\ mg\ drug$

1.57 The body first reacts to hypothermia by shivering. Further temperature lowering results in unconsciousness and later, followed by death.

1.59 Methanol would make a more effective cold compress because its higher specific heat allows it to retain the heat longer.

1.61 $d_{brain} = \dfrac{1\ \cancel{lb}}{620\ mL} \left(\dfrac{453.6\ g}{1\ \cancel{lb}} \right) = 0.732\ g/mL$

$specific\ gravity_{brain} = \dfrac{d_{brain}}{d_{H_2O}} = \dfrac{0.732\ \cancel{g/mL}}{1\ \cancel{g/mL}} = 0.732$

1.63 (a) potential energy   (b) kinetic energy   (c) potential energy
(d) kinetic energy   (e) kinetic energy

1.65 Convert European car's fuel efficiency of 22 km/L into mi/gal, then compare:

$fuel\ efficiency_{European} = \dfrac{22\ \cancel{km}}{\cancel{L}} \left(\dfrac{1\ mi}{1.609\ \cancel{km}} \right) \left(\dfrac{3.785\ \cancel{L}}{1\ gal} \right) = 52\ mi/gal$

The European car is more fuel efficient by 12 miles per gallon

1.67 Shivering generates kinetic energy.

1.69 Convert each quantity into a common unit (grams): (a) is the largest and (d) is the smallest.

(a) 41 g   (b) $3 \times 10^3 \; \cancel{mg} \left(\dfrac{1 \; g}{1000 \; \cancel{mg}} \right) = 3 \; g$ (c) $8.2 \times 10^6 \; \cancel{\mu g} \left(\dfrac{1 \; g}{10^6 \; \cancel{\mu g}} \right) = 8.2 \; g$

(d) $4.1310 \times 10^{-8} \; \cancel{kg} \left(\dfrac{1000 \; g}{1 \; \cancel{kg}} \right) = 4.1310 \times 10^{-5} \; g$

1.71   travel time $= 1490 \; \cancel{mi} \left(\dfrac{1.609 \; \cancel{km}}{1 \; \cancel{mi}} \right) \left(\dfrac{1 \; hr}{220 \; \cancel{km}} \right) = 10.9 \; hr$

1.73 SH (water) = 1.000 cal/g · °C = 4.184 J/g · °C
SH (heavy water) = 4.217 J/g · °C
heat = SH x m x ΔT: According to the equation, the heat required to raise the temperature of a substance is directly proportional to the specific heat of that substance. Heavy water, having the higher specific heat, will require more energy to heat 10 g by 10°C.

1.75 1.00 mL of butter = 0.860 g and 1.00 mL of sand = 2.28 g

(a) $d_{mixture} = \dfrac{3.14 \; g \; mixture}{2.00 \; mL} = 1.57 \; g/mL$

(b)   First calculate the volumes of sand and butter, which is 1.60 mL

$V_{sand} = 1.00 \; \cancel{g \; sand} \left(\dfrac{1.00 \; mL \; sand}{2.28 \; \cancel{g \; sand}} \right) = 0.439 \; mL$

$V_{butter} = 1.00 \; \cancel{g \; butter} \left(\dfrac{1.00 \; mL \; butter}{0.860 \; \cancel{g \; butter}} \right) = 1.16 \; mL$

$d_{mixture} = \dfrac{2.00 \; g}{1.60 \; mL} = 1.25 \; g/mL$

1.77 The final answer will be reported to two significant digits because the temperature is reported in two significant figures (the least accurate of the quantities used in the calculation)

$SH_{unk} = \dfrac{heat}{m \times (T_2 - T_1)} = \dfrac{3.200 \; kcal}{(92.15 \; g)(45°C)} = 7.7 \times 10^{-4} \; kcal/g \cdot °C = 0.77 \; cal/g \cdot °C$

1.79 $T_2 = \dfrac{\text{heat}}{\text{SH} \times \text{m}} + T_1$

$T_2 = \dfrac{60.0\ \cancel{J}}{\left(10.0\ \cancel{g}\right)\left(1.339\ \cancel{J}/\cancel{g}\cdot^\circ C\right)} + 20.0^\circ C = 24.5^\circ C$

Chapter 2 Atoms

<u>2.1</u> (a) $NaClO_3$ (b) AlF_3

<u>2.3</u> (a) The element has 15 protons, making it phosphorus (P); its symbol is $^{31}_{15}P$

(b) The element has 86 protons, making it radon (Rn); its symbol is $^{222}_{86}Rn$

<u>2.5</u> The atomic number of iodine (I) is 53. The number of neutrons in each isotope is 125 - 53 = 72 for iodine-125 and 131 - 53 = 78 for iodine-131. The symbols for these two isotopes are $^{125}_{53}I$ and $^{131}_{53}I$.

<u>2.7</u> This element has 13 electrons and, therefore, 13 protons. The element with atomic number 13 is Aluminum (Al).

Lewis Structure for Aluminum: Al$\vdots$

<u>2.9</u> The ring is made of a mixture.

<u>2.11</u> (a) Sulfur   (b) Iron   (c) Hydrogen   (d) Potassium   (e) Silver   (f) Gold

<u>2.13</u> Given here is the element, its symbol, and its atomic number:
   (a) Americium (Am, 95)   (b) Berkelium (Bk, 97)   (c) Californium (Cf, 98)
   (d) Dubnium (Db, 105)   (e) Europium (Eu, 63)   (f) Francium (Fr, 87)
   (g) Gallium (Ga, 31)   (h) Germanium (Ge, 32)   (i) Hafnium (Hf, 72)
   (j) Hassium (Hs, 108)   (k) Holmium (Ho, 67)   (l) Lutetium (Lu, 71)
   (m) Magnesium (Mg, 12)   (n) Polonium (Po, 84)   (o) Rhenium Re, 75)
   (p) Ruthenium ((Ru, 44)   (q) Scandium (Sc, 21)   (r) Strontium (Sr, 38)
   (s) Ytterbium (Yb, 70), Terbium (Tb, 65), and Yttrium (Y, 39)
   (t) Thulium (Tm, 69)

<u>2.15</u> (a) K_2O (b) Na_3PO_4 (c) $LiNO_3$

2.17 (a) The law of conservation of mass states that matter can be neither created nor destroyed. Dalton's theory explains this because if all matter is made up of indestructible atoms, then any chemical reaction just changes the attachments between atoms and does not destroy the atoms themselves.

(b) The law of constant composition states that any compound is always made up of elements in the same proportion by mass. Dalton's theory explains this because molecules consist of tightly bound groups of atoms, each of which has a particular mass. Therefore, each element in a compound always constitutes a fixed proportion of the total mass.

2.19 No. CO and CO_2 are different compounds, and each obeys the law of constant composition for that particular compound.

2.21 The statement is true in the sense that the number of protons (the atomic number) determines the identity of the element.

2.23 (a) The element with 22 protons is titanium (Ti)
(b) The element with 76 protons is osmium (Os)
(c) The element with 34 protons is selenium (Se)
(d) The element with 94 protons is plutonium (Pu)

2.25 Each would still be the same element because the number of protons has not changed.

2.27 Radon (Rn) has an atomic number of 86, so each isotope has 86 protons. The number of neutrons is mass number - atomic number.
(a) Radon-210 has 210 - 86 = 124 neutrons
(b) Radon-218 has 218 - 86 = 132 neutrons
(c) Radon-222 has 222 - 86 = 136 neutrons

2.29 two more neutrons: tin-120
three more neutrons: tin-121
four more neutrons: tin 124

2.31 (a) An ion is an atom with an unequal number of protons and electrons.
(b) Isotopes are atoms with the same number of protons in their nuclei but a different number of neutrons.

2.33 Rounded to three significant figures, the calculated value is 12.0 amu. The value given in the Periodic Table is 12.011 amu.

$$\left(\frac{98.90}{100} \times 12.000 \text{ amu}\right) + \left(\frac{1.10}{100} \times 13.000 \text{ amu}\right) = 12.011 \text{ amu}$$

2.35 Carbon-11 has 6 protons, 6 electrons, and 5 neutrons

2.37 Americium-241 (Am) has atomic number 95. This isotope has 91 protons, 91 electrons and 241 - 95 = 146 neutrons.

2.39 (a) Groups 2A, 3B, 4B, 5B, 6B, 7B, 8B, 1B, and 2B contain only metals. Note that Group 1A contains one nonmetal, hydrogen.
(b) No group contains only metalloids.
(c) Only Groups 7A and 8A contain only nonmetals.

2.41 Elements in the same group in the Periodic Table should have similar properties: As, N, and P; I and F; Ne and He; Mg, Ca, and Ba; K and Li.

2.43 (a) Aluminum > silicon   (b) Arsenic > phosphorus
(c) Gallium > germanium   (d) Gallium > aluminum

2.45 These numbers arise because (1) an orbital can hold two electrons, (2) s orbitals occur singly and can hold two electrons, (3) p orbitals occur in sets of three and can hold 6 electrons, (4) d orbitals occur in sets of five and can hold ten electrons, and f orbitals come in sets of seven and can hole 14 electrons. The first shell has only an s orbital and, therefore, can hold only two electrons. The second shell has $2s$ and $2p$ orbitals and can hold 8 electrons. The third shell has $3s$, $3p$, and $3d$ orbitals and can hold 18 electrons. The fourth shell has $4s$, $4p$, $4d$, and $4f$ orbitals, and can hold 32 electrons.

2.47 The group number tells the number of electrons in the valence shell of the element.

2.49 (a) Li(3): $1s^22s^1$ (b) Ne(10): $1s^22s^22p^6$ (c) Be(4): $1s^22s^2$
(d) C(6): $1s^22s^22p^2$ (e) Mg(12): $1s^22s^22p^63s^2$

2.51 (a) He(2): $1s^2$ (b) Na(11): $1s^22s^22p^63s^1$ (c) Cl(17): $1s^22s^22p^63s^23p^5$
(d) P(15): $1s^22s^22p^63s^23p^3$ (e) H(1): $1s^1$

2.53 In (a), (b), and (c); the outer-shell electron configurations are the same. The only difference is the number of the valence shell being filled.

9

2.55 The element might be in Group 2A, all of which have two valence electrons. It might also be helium.

2.57 The total is $(6s = 2) + (6p = 6) + (6d = 10) + (6f = 14) = 32$ electrons.

2.59 The properties are similar because all of them have the same outer-shell electron configuration. They are not identical because each has a different number of filled inner shells.

2.61 Sulfur and iron are essential components of proteins, and calcium is a major component of bones and teeth.

2.63 Because the $^2H/^1H$ ratio on Mars is five times larger than on Earth, the atomic weight of hydrogen on Mars would be greater than that on Earth.

2.65 Bronze is an alloy of copper and tin.

2.67 (a) $1s$ (b) $2s, 2p$ (c) $3s, 3p, 3d$ (d) $4s, 4p, 4d, 4f$

2.69 (a) s^2p^1 (b) s^2p^5 (c) s^2p^3

2.71 (a) Phosporous-32 has 15 protons, 15 electrons, and 32 - 15 = 17 neutrons.
(b) Molybdenum-98 has 42 protons, 42 electrons, and 98 - 42 = 56 neutrons.
(c) Calcium-44 has 20 protons, 20 electrons, and 44 - 20 = 24 neutrons.
(d) Hydrogen-3 has 1 proton, 1 electron, and 3 - 1 = 2 neutrons.
(e) Gadolinium-158 has 64 protons, 64 electrons, and 158 - 64 = 94 neutrons.
(f) Bismuth-212 has 83 protons, 83 electrons, and 212 - 83 = 129 neutrons.

2.73 Isotopes of elements from 37 to 53 contain more neutrons than protons.

2.75 Rounded to three significant figures, the atomic weight of naturally occurring boron is 10.8. The value given in the Periodic Table is 10.811.
$$\left(\frac{19.9}{100} \times 10.013 \text{ amu}\right) + \left(\frac{80.1}{100} \times 11.009 \text{ amu}\right) = 10.811 \text{ amu}$$

2.77 It would take 6.0×10^{21} protons to equal the mass of a grain of salt.
$$\frac{1.0 \times 10^{-2} \text{ g NaCl}}{1.67 \times 10^{-24} \text{ g/ proton}} = 6.0 \times 10^{21} \text{ protons}$$

<u>2.79</u> Assume the isotope mass is equal to the isotope mass number. This problem can be solved by using these relationships.

$$\frac{\%\,^{85}Rb}{100} + \frac{\%\,^{87}Rb}{100} = 1 \qquad \text{or} \qquad \frac{\%\,^{85}Rb}{100} = 1 - \frac{\%\,^{87}Rb}{100}$$

$$\left(\frac{\%\,^{85}Rb}{100} \times 85\right) + \left(\frac{\%\,^{87}Rb}{100} \times 87\right) = 85.47$$

$$\left[\left(1 - \frac{\%\,^{87}Rb}{100}\right) \times 85\right] + \left(\frac{\%\,^{87}Rb}{100} \times 87\right) = 85.47$$

$$85 - \left(\frac{\%\,^{87}Rb}{100} \times 85\right) + \left(\frac{\%\,^{87}Rb}{100} \times 87\right) = 85.47$$

$$2 \times \frac{\%\,^{85}Rb}{100} = 85.47 - 85$$

$$^{87}Rb = 23.5\%$$

$$^{85}Rb = 100 - 23.5 = 76.5\%$$

<u>2.81</u> Xenon (Xe) will have the highest ionization energy. Ionization energy increases from left to right going across the periodic table.

Chapter 3 Chemical Bonds

3.1 (a) Magnesium (Mg) atom with two valance electrons, loses both electrons to form a Mg^{2+} ion with a Neon (Ne) electron configuration.
(b) Sulfur (S) atom with six valance electrons gains two electrons to give a sulfide ion (S^{2-}) with eight valance electron octet of an argon electron configuration.

3.3 (a) KCl   (b) CaF_2 (c) Fe_2O_3

3.5 (a) $MgCl_2$ (b) Al_2O_3 (c) LiI

3.7 (a) Potassium hydrogen phosphate   (b) Aluminum sulfate
(c) Iron (II) carbonate or Ferrous carbonate

3.9 (a) $\overset{\delta+ \quad \delta-}{C - N}$ (b) $\overset{\delta+ \quad \delta-}{N - O}$ (c) $\overset{\delta+ \quad \delta-}{C - Cl}$

3.11 (a) Nitrogen dioxide   (b) Phosphorus tribromide   (c) Sulfur dichloride

3.13

(a) [structure of CH₃Cl with dipole arrow]
polar

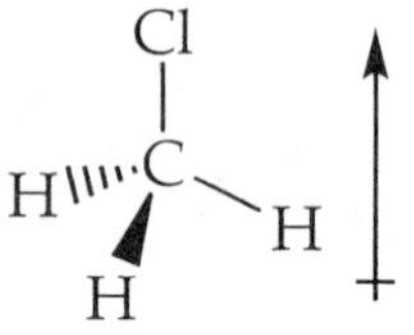

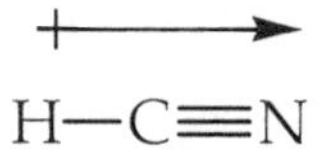

(b) H—C≡N
polar

(c) [structure of C₂H₆]
nonpolar

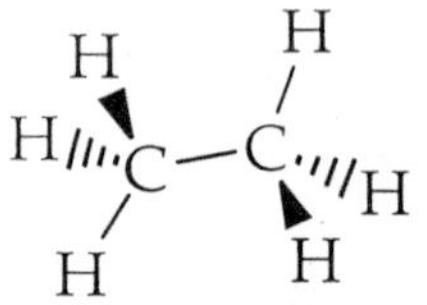

3.15 (a) Li: $1s^2 2s^1 \rightarrow Li^+$: $1s^2$ + 1 electron
(b) O: $1s^2 2s^2 2p^4$ + 2 electrons $\rightarrow O^{2-}$: $1s^2 2s^2 2p^6$ (complete octet)

3.17 (a) Mg^{2+} (b) F^- (c) Al^{3+} (d) S^{2-} (e) K^+ (f) Br^-

3.19 Li^- is not stable because it has an unfilled 2nd shell.

3.21 No. Copper is a transition metal so the octet rule does not apply. Transition metals can expand their octet into the $3d$-orbitals.

3.23 Sodium chloride in the solid state has is a Na^+ ion surrounded by six Cl^- anions and each Cl^- anion surrounded by six Na^+ ions.

3.25

	Br⁻	ClO₄⁻	O²⁻	NO₃⁻	SO₄²⁻	PO₄³⁻	OH⁻
Li⁺	LiBr	LiClO₄	Li₂O	LiNO₃	Li₂SO₄	Li₃PO₄	LiOH
Ca²⁺	CaBr₂	Ca(ClO₄)₂	CaO	Ca(NO₃)₂	CaSO₄	Ca₃(PO₄)₂	Ca(OH)₂
Co³⁺	CoBr₃	Co(ClO₄)₃	Co₂O₃	Co(NO₃)₃	Co₂(SO₄)₃	CoPO₄	Co(OH)₃
K⁺	KBr	KClO₄	K₂O	KNO₃	K₂SO₄	K₃PO₄	KOH
Cu²⁺	CuBr₂	Cu(ClO₄)₂	CuO	Cu(NO₃)₂	CuSO₄	Cu₃(PO₄)₂	Cu(OH)₂

3.27 (a) Hydrogen carbonate ion (bicarbonate ion) (b) Nitrite ion
(c) Sulfate ion (d) Hydrogen sulfate ion (e) Dihydrogen phosphate ion

3.29 The formula for potassium nitrate is KNO₂

3.31 (a) Na⁺, Br⁻ (b) Fe²⁺, SO₃²⁻ (c) Mg²⁺, PO₄³⁻
(d) K⁺, H₂PO₄⁻ (e) Na⁺, HCO₃⁻ (f) Ba²⁺, NO₃⁻

3.33 (a) Fe(OH)₃ (b) BaCl₂ (c) Ca₃(PO₄)₂ (d) NaMnO₄

3.35 (a) The formula, (NH₄)₂PO₄, is incorrect. The correct formula is: (NH₄)₃PO₄.
(b) The formula, Ba₂CO₃, is incorrect. The correct formula is BaCO₃.
(c) The formula for aluminum sulfide, Al₂O₃ is correct.
(d) The formula for magnesium sulfide, MgS is correct.

3.37 (a) KBr (b) CaO (c) HgO
(d) Cu₃(PO₄)₂ (e) Li₂SO₄ (e) Fe₂S₃

3.39 (a) Sodium fluoride (b) Magnesium sulfide (c) Aluminum oxide
(d) Barium chloride (e) Calcium hydrogen sulfite (calcium bisulfite)
(f) Potassium iodide (g) Strontium phosphate
(h) Iron(II) hydroxide (ferrous hydroxide) (i) Sodium dihydrogen phosphate
(j) Lead(II) acetate (plumbous acetate) (k) Barium hydride
(l) Ammonium hydrogen phosphate

3.41 Carbon forms up to four covalent bonds (e.g. methane, CH₄). Nitrogen usually forms up to three bonds (e.g. ammonia, NH₃) and up to four covalent bonds when positively charged (e.g. ammonium, NH₄⁺). Oxygen usually forms up to two covalent bonds (e.g. H₂O) and sometimes three covalent bonds when positively charged (e.g. hydronium, H₃O⁺). Fluorine and bromine form one covalent bond (e.g. HBr and HF).

Chapter 3 Chemical Bonds

<u>3.43</u> The following seven diatomic elements are hydrogen, nitrogen, oxygen, fluorine, chlorine, bromine, and iodine.

(a)  H–H   :N≡N: $\overset{..}{\underset{..}{O}}=\overset{..}{\underset{..}{O}}$ $:\overset{..}{\underset{..}{F}}-\overset{..}{\underset{..}{F}}:$ $:\overset{..}{\underset{..}{Cl}}-\overset{..}{\underset{..}{Cl}}:$ $:\overset{..}{\underset{..}{Br}}-\overset{..}{\underset{..}{Br}}:$ $:\overset{..}{\underset{..}{I}}-\overset{..}{\underset{..}{I}}:$

(b) gases: H_2, N_2, O_2, F_2, Cl_2; liquids: Br_2; solids: I_2

<u>3.45</u> A molecular formula tells us the number of atoms in the simplest unit of a compound. A structural formula shows how the atoms are bonded in a molecule. A Lewis structure shows how the atoms are bonded in a molecule as well as the location of unshared electrons.

<u>3.47</u> Total number of valence electrons for each compound:
(a) NH_3 has 8   (b) C_3H_6 has 18   (c) $C_2H_4O_2$ has 24
(d) C_2H_6O has 20   (e) CCl_4 has 32   (f) HNO_2 has 18
(g) CCl_2F_2 has 32   (h) O_2 has 12

<u>3.49</u> A bromine atom contains seven electrons in its valence shell. A bromine molecule contains two bromine atoms bonded by a single bond. A bromide ion is a bromine atom that has gained one electron in its valence shell; it has a complete octet and a charge of -1.

(a) 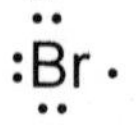 $:\overset{..}{Br}\cdot$ (b) 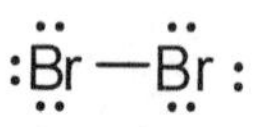 $:\overset{..}{\underset{..}{Br}}-\overset{..}{\underset{..}{Br}}:$ (c) 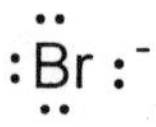 $:\overset{..}{\underset{..}{Br}}:^-$

<u>3.51</u> (a) Incorrect. The left carbon has five bonds.
(b) Incorrect. The carbon in the middle has only three bonds.
(c) Incorrect. The second carbon from the right has only three bonds and the oxygen on the right has only one bond.
(d) Incorrect. Fluorine has two bonds.
(e) Correct.
(f) Incorrect. The second carbon from the left has five bonds.

<u>3.53</u> (a) H_2O has 8 valence electrons, and H_2O_2 has 14 valence electrons.

(b) 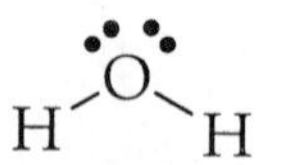 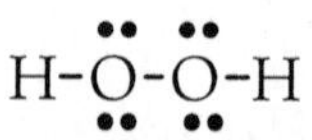 $H\overset{\overset{..}{O}}{\diagup}{}_{\diagdown}H$ $H-\overset{..}{\underset{..}{O}}-\overset{..}{\underset{..}{O}}-H$

(c) Predicted bond angles of 109.5° about each oxygen atom.

14

3.55 Under each Lewis structure is the number of valence electrons in the molecule and the predicted bond angles about nitrogen.

$$H-\ddot{N}-H \qquad H-\ddot{N}-\ddot{N}-H \qquad H-\ddot{N}=\ddot{N}-H$$
$$\quad\ \ \ | \qquad\qquad\quad\ | \ \ | $$
$$\quad\ \ \ H \qquad\qquad\quad\ H\ H$$

Ammonia	Hydrazine	Diimide
8 valence electrons	14 valence electrons	12 valence electrons
109.5°	109.5°	120°

3.57 Shape of the each molecule and approximate bond angles about its central atom:
(a) Tetrahedral, 109.5° (b) Pyramidal, 109.5° (c) Tetrahedral, 109.5
(d) Bent, 120° (e) Trigonal planar, 120° (f) Tetrahedral, 109.5°
(g) Bent, 120° (h) Pyramidal, 109.5° (i) Tetrahedral, 109.5°
(j) Pyramidal, 109.5°

3.59 Electronegativity generally increases in going from left to right across a row because the number of positive charges in the nucleus of each element in a row increases going from left to right. The increasing nuclear charge exerts a stronger and stronger pull on the valence electrons.

3.61 Electrons are shifted toward the more electronegative atom.
(a) Cl   (b) O   (c) O   (d) Cl   (e) C   (f) S   (g) O

3.63 The differences are in their shapes. Because CO_2 is a linear molecule, it is nonpolar. SO_2 is a bent molecule, therefore it is polar.

(a) $\ddot{O}=C=\ddot{O}$ (b)

3.65 No, molecular dipoles are the result of the sum of the direction and magnitude of each individual polar bond.

3.67 For molecules where a bond will form, use differences in electronegativity to predict the polarity of the bond.
(a) Cl-I (3.0 - 2.5 = 0.5); a polar covalent bond.
(b) No bond. Each atom loses rather than accept electrons.
(c) K-O (3.5 - 0.8 = 2.7); an ionic bond.
(d) Al-N (3.0 - 1.5 = 1.5); a polar covalent bond.
(e) No bond. Each atom loses rather than accept electrons.
(f) No bond. Neon is a noble gas and neither gains, loses, or shares electrons.
(g) No bond. Each atom loses rather than accept electrons.
(h) No bond. Helium is a noble gas and neither gains, loses, or shares electrons.

3.69 Titanium dioxide, TiO_2, is used in sunblock to reflect sunlight from the skin.

3.71 Lead(IV) oxide, PbO_2, and lead(IV) carbonate, $Pb(CO_3)_2$, were used as pigments in paint.

3.73 Fe(II) is utilized in over-the-counter iron supplements.

3.75 (a) $CaSO_3$ (b) $Ca(HSO_3)_2$ (c) $Ca(OH)_2$ (d) $CaHPO_4$

3.77 Calcium dihydrogen phosphate, calcium phosphate, and calcium carbonate.

3.79 Barium sulfate is used to visualize the gastrointestinal tract by x-ray examination.

3.81 Calcium (Ca^{2+}) is the main metal ion present in bone and tooth enamel

3.83 Hydrogen has the electron configuration $1s^1$. Hydrogen's valence shell has only a $1s$ orbital, which can hold only two electrons.

3.85 When carbon has four bonds, it is surrounded by eight electrons and has a complete octet. An additional lone pair of electrons would exceed carbon's octet.

3.87 Nitrogen has five valence electrons. By sharing three more electrons with another atom(s), nitrogen can achieve the outer-shell electron configuration of neon, the noble gas nearest to it in atomic number. The three shared pairs of electrons may be in the form of three single bonds, one double bond and one single bond, or one triple bond. With any of these combinations, there is one unshared pair of electrons on nitrogen.

3.89 Oxygen has six valence electrons. By sharing two electrons with another atom(s), oxygen can achieve the outer-shell electron configuration of neon, the Noble gas nearest to it in atomic number. The two shared pairs of electrons may be in the form of a double bond or two single bonds. With either of these configurations, there are two unshared pairs of electrons on the oxygen.

3.91 O^{6+} has a charge too concentrated for a small ion.

3.93 Use the difference in electronegativity to determine the character of the bond.
(a) C-Cl (3.0 - 2.5 = 0.5); polar covalent   (b) C-Li (2.5 - 1.0 = 1.5); polar covalent
(c) C-N (3.0 - 2.5 = 0.5); polar covalent

3.95 The problem is not with carbon forming a double bond to chlorine, but with chlorine forming a double bond to anything. Because chlorine needs only one electron to complete its valence shell, it forms only a single covalent bond with another element.

3.97 (a) BF_3 has six valence electrons around the boron, thus does not obey octet rule.
(b) CF_2: does not obey the octet rule because carbon has 4 electrons around it.
(c) BeF_2: does not obey the octet rule because Be has 4 electrons around it.
(d) $H_2C=CH_2$ obeys the octet rule
(e) CH_3 does not obey the octet rule because carbon has 6 electrons around it.
(f) N_2 obeys the octet rule.
(g) NO does not obey the octet rule Lewis Structures drawn for the compound shows either a nitrogen or an oxygen atom with 7 electrons.

3.99 Predict a shape like that created by putting together the bases of two square-based pyramids. This shape is called octahedral, because it has eight sides.

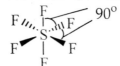

The VSPER theory would predict an octahedral structure, which places the bonding pairs of electrons as far apart as possible, thereby minimizing electron pair repulsions.

3.101 (a) $SiCl_4$ (b) PH_3 (c) H_2S

3.103 (a) Following is a Lewis structure for vinyl chloride.
(b) All bond angles are predicted to be 120°.
(c) Vinyl chloride has a polar C-Cl bond, is a polar molecule, and has a dipole.

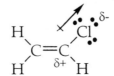

3.105 (a) ClO_2 has 19 valence electrons. In the Lewis structure, either the chlorine or one of the oxygens must have only seven valence electrons. In the following structure, the odd electron is placed on the chlorine, which is the least electronegative element of the two atoms.
(b) Lewis structure of ClO_2:

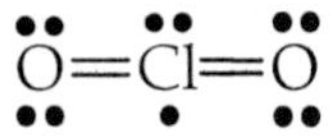

17

Chapter 4 Chemical Reactions

<u>4.1</u> (a) Aspirin, $C_9H_8O_4$

C 9×12.0 amu = 108 amu

H 8×1.00 amu =   8.00 amu

O 4×16.0 amu = 64.0 amu

$C_9H_8O_4$ = 180.0 amu

(b) Barium Phosphate, $Ba_3(PO_4)_2$

Ba 3×137 amu = 411 amu

P 2×31.0 amu = 62.0 amu

O 8×16.0 amu = 128 amu

$Ba_3(PO_4)_2$ = 601 amu

<u>4.3</u> $2.84 \ \cancel{mol \ Na_2S} \left(\dfrac{87.1 \text{ g } Na_2S}{1 \ \cancel{mol \ Na_2S}} \right) = 222 \text{ g } Na_2S$

<u>4.5</u> moles of Cu(I) ions:

$$0.062 \ \cancel{\text{g } CuNO_3} \left(\frac{1 \ \cancel{\text{mol } CuNO_3}}{125.5 \ \cancel{\text{g } CuNO_3}} \right) \left(\frac{1 \text{ mol } Cu^+ \text{ ions}}{1 \ \cancel{\text{mol } CuNO_3}} \right) = 4.9 \times 10^{-4} \text{ mol } Cu^+$$

<u>4.7</u> Balance: $CO_2(g) + H_2O(l) \longrightarrow C_6H_{12}O_6(aq) + O_2(g)$

step 1: balance carbons with a coefficient of 6 in front of CO_2

$6CO_2(g) + H_2O(l) \longrightarrow C_6H_{12}O_6(aq) + O_2(g)$

step 2: balance hydrogen with a coefficient of 6 in front of H_2O

$6CO_2(g) + 6H_2O(l) \longrightarrow C_6H_{12}O_6(aq) + O_2(g)$

step 3: last step, balance oxygen with a coefficient of 6 in front of O_2

balanced equation: $6CO_2(g) + 6H_2O(l) \xrightarrow{\text{photosynthesis}} C_6H_{12}O_6(aq) + 6O_2(g)$

<u>4.9</u> Balance the equation: $K_2C_2O_4(aq) + Ca_3(AsO_4)_2(s) \longrightarrow K_3AsO_4(aq) + CaC_2O_4(s)$

step 1: first balance the most complicated AsO4 with a 2 in front of K_3AsO_4

$K_2C_2O_4(aq) + Ca_3(AsO_4)_2(s) \longrightarrow 2K_3AsO_4(aq) + CaC_2O_4(s)$

step 2: next balance the potassium by placing a 3 in front of $K_2C_2O_4$

$3K_2C_2O_4(aq) + Ca_3(AsO_4)_2(s) \longrightarrow 2K_3AsO_4(aq) + CaC_2O_4(s)$

step 3: now balance C_2O_4 and Ca with a coefficient of 3 in front of CaC_2O_4

balanced equation:

$3K_2C_2O_4(aq) + Ca_3(AsO_4)_2(s) \longrightarrow 2K_3AsO_4(aq) + 3CaC_2O_4(s)$

<u>4.11</u> From the balanced equation: $CH_3OH(g) + CO(g) \longrightarrow CH_3COOH(l)$,

The molar ratio of CO required to produce CH_3COOH is 1:1. therefore 16.6 moles of CO is required to produce 16.6 moles of CH_3COOH.

2

Chapter 4 Chemical Reactions

<u>4.13</u> 6.0 g carbon = 0.50 mol of carbon   2.1 g H_2 = 1.1 mol H_2

mass of H_2 required = 6.0 g C $\left(\dfrac{1\ \text{mol C}}{12.0\ \text{g C}}\right)\left(\dfrac{2\ \text{mol }H_2}{1\ \text{mol C}}\right)\left(\dfrac{2.0\ \text{g}}{1\ \text{mol }H_2}\right)$ = 2.0 g H_2

mass of C required = 2.1 g H_2 $\left(\dfrac{1\ \text{mol }H_2}{2.0\ \text{g }H_2}\right)\left(\dfrac{1\ \text{mol C}}{2\ \text{mol }H_2}\right)\left(\dfrac{12.0\ \text{g C}}{1\ \text{mol C}}\right)$ = 6.2 g C

(a) H_2 is in excess and C is the limiting reagent.
(b) 8.0 g CH_4 are produced.

6.0 g C $\left(\dfrac{1\ \text{mol C}}{12.0\ \text{g C}}\right)\left(\dfrac{1\ \text{mol }CH_4}{1\ \text{mol C}}\right)\left(\dfrac{16.0\ \text{g }CH_4}{1\ \text{mol }CH_4}\right)$ = 8.0 g CH_4 produced

<u>4.15</u> Overall chemical reaction: $CuCl_2(aq)$ + $K_2S(aq)$ → $CuS(s)$ + $2KCl(aq)$
step 1: write a equation involving all of the chemical species participating in the chemical reaction.
$Cu^{2+}(aq)$ + $2Cl^-(aq)$ + $2K^+(aq)$ + $S^{2-}(aq)$ → $CuS(s)$ + $2K^+(aq)$ + $2Cl^-(aq)$
step 2: cross out the aqueous ions that appear on both sides of the equation.
$Cu^{2+}(aq)$ + $2Cl^-(aq)$ + $2K^+(aq)$ + $S^{2-}(aq)$ → $CuS(s)$ + $2K^+(aq)$ + $2Cl^-(aq)$

net ionic equation: $Cu^{2+}(aq)$ + $S^{2-}(aq)$ → $CuS(s)$

The net ionic equation shows the chemical species that actually undergo a chemical change. The ions that appear on both sides of the equation do not change, therefore are considered spectator ions.

<u>4.17</u> (a) Cl_2 = 71.0 amu   (b) Ar = 39.9 amu   (c) P_4 = 124.0 amu
(d) N_2 = 28.0 amu   (e) He = 4.0 amu

<u>4.19</u> (a) 32 g CH_4 $\left(\dfrac{1\ \text{mole }CH_4}{16.0\ \text{g }CH_4}\right)$ = 2.0 mol CH_4

(b) 345.6 g NO $\left(\dfrac{1\ \text{mol NO}}{30.01\ \text{g NO}}\right)$ = 11.52 mol NO

(c) 184.4 g ClO_2 $\left(\dfrac{1\ \text{mol }ClO_2}{67.45\ \text{g }ClO_2}\right)$ = 2.754 mol ClO_2

(d) 720 g $C_3H_8O_3$ $\left(\dfrac{1\ \text{mol }C_3H_8O_3}{92.1\ \text{g }C_3H_8O_3}\right)$ = 7.82 mol $C_3H_8O_3$

Chapter 4 Chemical Reactions

4.21 (a) $18.1 \text{ mol CH}_2\text{O} \left(\dfrac{1 \text{ mol O atoms}}{1 \text{ mol CH}_2\text{O}} \right) = 18.1$ mol O atoms

(b) $0.41 \text{ mol CHBr}_3 \left(\dfrac{3 \text{ mol Br atoms}}{1 \text{ mol CHBr}_3} \right) = 1.2$ mol Br atoms

(c) $3.5 \times 10^3 \text{ mol Al}_2(\text{SO}_4)_3 \left(\dfrac{12 \text{ O atoms}}{1 \text{ mol Al}_2(\text{SO}_4)_3} \right) = 4.2 \times 10^4$ mol O atoms

(d) $87 \text{ g HgO} \left(\dfrac{1 \text{ mol HgO}}{216.6 \text{ g HgO}} \right) \left(\dfrac{1 \text{ mol Hg atoms}}{1 \text{ mol HgO}} \right) = 0.40$ mol Hg atoms

4.23 The same; that is, just 2:1.

4.25 $3.1 \times 10^{-1} \text{ g C}_9\text{H}_8\text{O}_4 \left(\dfrac{1 \text{ mol C}_9\text{H}_8\text{O}_4}{180 \text{ g C}_9\text{H}_8\text{O}_4} \right) \left(\dfrac{6.02 \times 10^{23} \text{ molec}}{1 \text{ mol C}_9\text{H}_8\text{O}_4} \right) = 1.0 \times 10^{21}$ molec $C_9H_8O_4$

4.27 $3.9 \text{ mg chol.} \left(\dfrac{1 \text{ g chol.}}{1000 \text{ mg chol.}} \right) \left(\dfrac{1 \text{ mol chol.}}{386.7 \text{ g chol.}} \right) = 1.0 \times 10^{-5}$ mol cholesterol

$1.0 \times 10^{-5} \text{ mol chol.} \left(\dfrac{6.02 \times 10^{23} \text{ molec. chol.}}{1 \text{ mol chol.}} \right) = 6.1 \times 10^{18}$ molecules of cholesterol

4.29 The following are the balanced equations:
(a) $HI + NaOH \rightarrow NaI + H_2O$
(b) $Ba(NO_3)_2 + H_2S \rightarrow BaS + 2HNO_3$
(c) $CH_4 + O_2 \rightarrow CO_2 + 2H_2O$
(d) $2C_4H_{10} \ 13O_2 \rightarrow 8CO_2 + 10H_2O$
(e) $2Fe + 3CO_2 \rightarrow Fe_2O_3 + 3CO$

4.31 $CO_2(g) + Ca(OH)_2(aq) \rightarrow CaCO_3(s) + H_2O(l)$

4.33 $2Mg(s) + O_2(g) \rightarrow 2MgO(s)$

4.35 $N_2O_5(g) + H_2O(l) \rightarrow 2HNO_3(aq)$

4.37 $(NH_4)_2CO_3(s) \rightarrow 2NH_3(g) + CO_2(g) + H_2O(l)$

Chapter 4 Chemical Reactions

<u>4.39</u> $2Al(s) + 3HCl(aq) \rightarrow 2AlCl_3(aq) + 3H_2(g)$

<u>4.41</u> From the balanced equation: $2N_2(g) + 3O_2(g) \rightarrow 2N_2O_3(g)$

(a) $1 \text{ mol } O_2 \left(\dfrac{2 \text{ mol } N_2}{3 \text{ mol } O_2} \right) = 0.67 \text{ mol } N_2$ required

(b) $1 \text{ mol } O_2 \left(\dfrac{2 \text{ } N_2O_3}{3 \text{ mol } O_2} \right) = 0.67 \text{ mol } N_2O_3$ produced

(c) $8 \text{ mol } N_2O_3 \left(\dfrac{3 \text{ mol } O_2}{2 \text{ mol } N_2O_3} \right) = 12 \text{ mol } O_2$ required

<u>4.43</u> From the balanced equation: $CH_4(g) + 3Cl_2(l) \rightarrow CHCl_3(g) + 3HCl(g)$

$1.50 \text{ mol } CHCl_3 \left(\dfrac{3 \text{ mol } Cl_2}{1 \text{ mol } CHCl_3} \right)\left(\dfrac{70.9 \text{ g } Cl_2}{1 \text{ mol } Cl_2} \right) = 319 \text{ g } Cl_2$ needed

<u>4.45</u> (a) balanced equation: $2NaClO_2(aq) + Cl_2(g) \rightarrow 2ClO_2(g) + 2NaCl(aq)$
(b) 4.1 kg ClO₂

$5.50 \text{ kg } NaClO_2 \left(\dfrac{1000 \text{ g } NaClO_2}{1 \text{ kg } NaClO_2} \right)\left(\dfrac{1 \text{ mol } NaClO_2}{90.4 \text{ g } NaClO_2} \right) = 60.8 \text{ mol } NaClO_2$

$60.8 \text{ mol } NaClO_2 \left(\dfrac{2 \text{ mol } ClO_2}{2 \text{ mol } NaClO_2} \right)\left(\dfrac{67.5 \text{ g } ClO_2}{1 \text{ mol } ClO_2} \right)\left(\dfrac{1 \text{ kg}}{1000 \text{ g}} \right) = 4.10 \text{ kg } ClO_2$

<u>4.47</u> From the balanced equation: $6CO_2(g) + 6H_2O(l) \rightarrow C_6H_{12}O_6(aq) + 6O_2(g)$

$5.1 \text{ g glucose} \left(\dfrac{1 \text{ mol glucose}}{180 \text{ g glucose}} \right)\left(\dfrac{6 \text{ mol } CO_2}{1 \text{ mol glucose}} \right)\left(\dfrac{44.0 \text{ g } CO_2}{1 \text{ mol } CO_2} \right) = 7.5 \text{ g } CO_2$

<u>4.49</u> From the balanced equation in problem #4.48

$0.58 \text{ g } Fe_2O_3 \left(\dfrac{1 \text{ mol } Fe_2O_3}{159.7 \text{ g } Fe_2O_3} \right)\left(\dfrac{6 \text{ mol } C}{2 \text{ mol } Fe_2O_3} \right)\left(\dfrac{12.0 \text{ g } C}{1 \text{ mol } C} \right) = 0.13 \text{ g C needed}$

4.51 From the balanced equation: $C_6H_6(l) + Br_2(l) \rightarrow C_6H_5Br(l) + HBr(g)$

$$\text{mass of } C_6H_6 \text{ required} = 135 \text{ g Br}_2 \left(\frac{1 \text{ mol Br}_2}{159.8 \text{ g Br}_2} \right)\left(\frac{1 \text{ mol } C_6H_6}{1 \text{ mol Br}_2} \right)\left(\frac{78.1 \text{ g } C_6H_6}{1 \text{ mol } C_6H_6} \right) = 66.0 \text{ g } C_6H_6$$

$$\text{mass of } Br_2 \text{ required} = 60.0 \text{ g } C_6H_6 \left(\frac{1 \text{ mol } C_6H_6}{78.1 \text{ g } C_6H_6} \right)\left(\frac{1 \text{ mol Br}_2}{1 \text{ mol } C_6H_6} \right)\left(\frac{159.8 \text{ g Br}_2}{1 \text{ mol Br}_2} \right) = 123 \text{ g Br}_2$$

(a) If 135 grams of bromine were used in the reaction, 66.0 grams of benzene would be required, which is more than the 60 grams of benzene available. Therefore, <u>benzene is the limiting reagent and the bromine is in excess.</u>

(b) $60.0 \text{ g } C_6H_6 \left(\frac{1 \text{ mol } C_6H_6}{78.1 \text{ g } C_6H_6} \right)\left(\frac{1 \text{ mol } C_6H_5Br}{1 \text{ mol } C_6H_6} \right)\left(\frac{157.0 \text{ g } C_6H_5Br}{1 \text{ mol } C_6H_5Br} \right) = 121 \text{ g } C_6H_5Br$

4.53 From the balanced equation: $CH_3CH_3(g) + Cl_2(g) \rightarrow CH_3CH_2Cl(l) + HCl(g)$
theoretical yield of ethyl chloride:

$$5.6 \text{ g ethane} \left(\frac{1 \text{ mol ethane}}{30.1 \text{ g ethane}} \right)\left(\frac{1 \text{ mol } CH_3CH_2Cl}{1 \text{ mol ethane}} \right)\left(\frac{64.5 \text{ g } CH_3CH_2Cl}{1 \text{ mol } CH_3CH_2Cl} \right) = 12 \text{ g}$$

actual yield of $CH_3CH_2Cl = 8.2$ g $\qquad$ % yield $= \dfrac{\text{actual yield}}{\text{theoretical yield}} \times 100$

$$\text{\% yield of } CH_3CH_2Cl = \frac{8.2 \text{ g } CH_3CH_2Cl}{12 \text{ g } CH_3CH_2Cl} \times 100 = 68\%$$

4.55 (a) Spectator ion: an ion that does not take part in a chemical reaction
(b) Net ionic equation: a balanced equation showing only the ions that react
(c) Aqueous solution: a solution using water as a solvent

4.57 (a) The spectator ions are Na^+ and Cl^-.
(b) $Na^+(aq) + CO_3^{2-}(aq) + Sr^{2+}(aq) + Cl^-(aq) \rightarrow SrCO_3(s) + Na^+(aq) + Cl^-(aq)$
$\qquad$ balanced net ionic equation: $CO_3^{2-}(aq) + Sr^{2+}(aq) \rightarrow SrCO_3(s)$

4.59 $Pb^{2+}(aq) + 2NO_3^-(aq) + 2NH_4^+(aq) + 2Cl^-(aq) \rightarrow$

$$PbCl_2(s) + 2NO_3^-(aq) + 2NH_4^+(aq)$$

$\qquad$ balanced net ionic equation: $Pb^{2+}(aq) + 2Cl^-(aq) \rightarrow PbCl_2(s)$

4.61 $2Na^+(aq)$ + $2OH^-(aq)$ + $2NH_4^+(aq)$ + $CO_3^{2-}(aq)$ →

$$2NH_3(g) + 2H_2O(l) + 2Na^+(aq) + CO_3^{2-}(aq)$$

balanced net ionic equation: $NH_4^+(aq) + OH^-(aq) \rightarrow NH_3(g) + H_2O(l)$

4.63   (a) $MgCl_2$ (soluble): most compounds containing Cl^- are soluble
      (b) $CaCO_3$ (insoluble): most compounds containing CO_3^{2-} are insoluble
      (c) Na_2SO_4 (soluble): all compounds containing Na^+ are soluble
      (d) NH_4NO_3 (soluble): all compounds containing NO_3^- and NH_4^+ are soluble
      (e) $Pb(OH)_2$ (insoluble): most compounds containing OH^- are insoluble

4.65   No, one species gains electrons and the other loses electrons. Electrons can not be destroyed, but transferred from one chemical species to another.

4.67   (a) C_7H_{12} is oxidized (the carbons gain oxygen going to CO_2) and O_2 is reduced.
      (b) O_2 is the oxidizing agent and C_7H_{12} is the reducing agent.

4.69   An exothermic chemical reaction or process releases heat as a product.
      An endothermic chemical reaction or process absorbs heat as a reactant.

4.71   19.6 kcal are given off.

4.73 $15.0 \text{ g glucose} \left(\dfrac{1 \text{ mol glucose}}{180 \text{ g glucose}} \right) \left(\dfrac{670 \text{ kcal}}{1 \text{ mol glucose}} \right) = 55.8$ kcal of heat evolved

4.75   (a) The synthesis of starch is endothermic.
      (b) 26.4 kcal

$$6.32 \text{ g starch} \left(\frac{10^{-3} \text{ kg starch}}{1 \text{ g starch}} \right) \left(\frac{4178 \text{ kcal}}{1.00 \text{ kg starch}} \right) = 26.4 \text{ kcal heat required}$$

4.77   From the balanced equation:

$$156.0 \text{ kcal} \left(\frac{2 \text{ mol Fe}}{196.5 \text{ kcal}} \right) \left(\frac{55.85 \text{ g Fe}}{1 \text{ mol Fe}} \right) = 88.68 \text{ g Fe metal produced}$$

4.79   Hydroxyapatite is composed of calcium ions, phosphate ions, and hydroxide ions.

4.81 C_2H_4O is oxidized (gain of oxygen) and H_2O_2 is reduced (loss of oxygen). H_2O_2 is the oxidizing agent and C_2H_4O is the reducing agent.

<u>4.83</u> (a) Fe_2O_3 loses oxygen; it is reduced. CO gains oxygen: it is oxidized.

(b) $38.4 \; \text{mol Fe} \left(\dfrac{1 \; \text{mol Fe}_2O_3}{2 \; \text{mol Fe}} \right) = 19.2 \; \text{mol Fe}_2O_3$ needed

(c) $38.4 \; \text{mol Fe} \left(\dfrac{3 \; \text{mol CO}}{2 \; \text{mol Fe}} \right)\left(\dfrac{28.01 \; \text{g CO}}{1 \; \text{mol CO}} \right) = 1.61 \times 10^3 \; \text{g CO}$ required

<u>4.85</u> The spectator ions are Na^+ and NO_3^-.

$6Na^+(aq) + 2PO_4^{3-}(aq) + 3Cd^{2+}(aq) + 6NO_3^-(aq) \rightarrow$

$$Cd_3(PO_4)_2(s) + 6NO_3^-(aq) + 6Na^+(aq)$$

balanced net ionic equation: $3Cd^{2+}(aq) + 2PO_4^{3-}(aq) \rightarrow Cd_3(PO_4)_2(s)$

<u>4.87</u> $MW_{chlorophyll} = \dfrac{24.305 \; \text{g Mg/mol}}{0.0272 \; \text{g Mg/1 g chlorophyll}} = 893 \; \text{amu}$

<u>4.89</u> 8.00 g $Pb(NO_3)_2$ added to 2.67 g $AlCl_3$ yielded 5.55g $PbCl_2$

mass of aluminum chloride required based on 8.00 g $Pb(NO_3)_2$:

$$8.00 \text{ g Pb(NO}_3)_2 \left(\frac{1 \text{ mol Pb(NO}_3)_2}{331.2 \text{ g Pb(NO}_3)_2} \right) \left(\frac{2 \text{ mol AlCl}_3}{3 \text{ mol Pb(NO}_3)_2} \right) = 1.61 \times 10^{-2} \text{ mol AlCl}_3$$

$$1.61 \times 10^{-2} \text{ mol AlCl}_3 \left(\frac{133.3 \text{ g AlCl}_3}{1 \text{ mol AlCl}_3} \right) = 2.15 \text{ g AlCl}_3 \text{ needed}$$

mass of lead(II) nitrate required based on 2.67 g $AlCl_3$:

$$2.67 \text{ g AlCl}_3 \left(\frac{1 \text{ mol AlCl}_3}{133.3 \text{ g AlCl}_3} \right) \left(\frac{3 \text{ mol Pb(NO}_3)_2}{2 \text{ mol AlCl}_3} \right) = 3.00 \times 10^{-2} \text{ mol of Pb(NO}_3)_2$$

$$3.00 \times 10^{-2} \text{ mol Pb(NO}_3)_2 \left(\frac{331.2 \text{ g Pb(NO}_3)_2}{1 \text{ mol Pb(NO}_3)_2} \right) = 9.94 \text{ g Pb(NO}_3)_2 \text{ needed}$$

(a) $Pb(NO_3)_2$ is the limiting reagent.

(b) actual yield of $PbCl_2$:

$$8.00 \text{ g Pb(NO}_3)_2 \left(\frac{1 \text{ mol Pb(NO}_3)_2}{331.2 \text{ g Pb(NO}_3)_2} \right) \left(\frac{3 \text{ mol PbCl}_2}{3 \text{ mol Pb(NO}_3)_2} \right) = 2.42 \times 10^{-2} \text{ mol PbCl}_2$$

$$2.42 \times 10^{-2} \text{ mol PbCl}_2 \left(\frac{278.1 \text{ g PbCl}_2}{1 \text{ mol PbCl}_2} \right) = 6.73 \text{ g PbCl}_2$$

$$\% \text{ yield} = \frac{\text{actual yield}}{\text{theoretical yield}} = \frac{5.55 \text{g PbCl}_2}{6.73 \text{ g PbCl}_2} \times 100 = 82.5\% \text{ PbCl}_2$$

<u>4.91</u> (a) $C_5H_{12}(g) + 8O_2(g) \rightarrow 5CO_2(g) + 6H_2O(g)$

(b) Pentane is oxidized and oxygen is reduced.

(c) Oxygen is the oxidizing agent and pentane is the reducing agent.

Chapter 5 Gases, Liquids, and Solids

5.1 $\quad P_2 = \dfrac{P_1 V_1}{V_2} = \dfrac{(0.70 \text{ atm})(3.8 \, \cancel{L})}{6.5 \, \cancel{L}} = 0.41 \text{ atm}$

5.3 $\quad P_2 = \dfrac{P_1 V_1 T_2}{T_1 V_2} = \dfrac{(0.92 \text{ atm})(20.5 \, \cancel{L})(285 \, \cancel{K})}{(296 \, \cancel{K})(340.6 \, \cancel{L})} = 0.053 \text{ atm}$

5.5 Ideal Gas Law: $PV = nRT$

$\quad n = \dfrac{PV}{RT} = \dfrac{(1.05 \, \cancel{\text{atm}})(10.0 \, \cancel{L})}{(0.0821 \, \cancel{L} \cdot \cancel{\text{atm}} \cdot \text{mol}^{-1} \cdot \cancel{K})(303 \, \cancel{K})} = 0.422 \text{ mol Ne}$

5.7 Dalton's Law of Partial Pressures:

total pressure $(P_T) = P_{N_2} + P_{H_2O}$

$P_{H_2O} = P_T - P_{N_2} = 2.015 \text{ atm} - 1.908 \text{ atm} = 0.107 \text{ atm of } H_2O \text{ vapor}$

5.9 heat of vaporization of water = 540 cal/g

$\quad 45.0 \, \cancel{\text{kcal}} \left(\dfrac{1000 \, \cancel{\text{cal}}}{1 \, \cancel{\text{kcal}}} \right) \left(\dfrac{1 \text{ g } H_2O}{540 \, \cancel{\text{cal}}} \right) = 83.3 \text{ g } H_2O \text{ vaporized}$

5.11 According to the phase diagram of water (figure 5.18), the vapor will undergo reverse sublimation and form a solid.

5.13 Boyle's Law:

at constant temperature, $\left(\dfrac{P_1 V_1}{\cancel{T_1}} \right) = \left(\dfrac{P_2 V_2}{\cancel{T_2}} \right)$ reduces to $P_1 V_1 = P_2 V_2$

$\quad P_1 = \dfrac{P_2 V_2}{V_1} = \dfrac{(12.2 \text{ atm})(2.5 \, \cancel{L})}{20 \, \cancel{L}} = 1.5 \text{ atm } CH_4$

5.15 Gay-Lussac's Law: The tire is at constant volume

at constant volume, $\left(\dfrac{P_1 \cancel{V_1}}{T_1} \right) = \left(\dfrac{P_2 \cancel{V_2}}{T_2} \right)$ reduces to $\dfrac{P_1}{T_1} = \dfrac{P_2}{T_2}$

$\quad P_2 = \dfrac{P_1 T_2}{T_1} = \dfrac{(2.30 \text{ atm})(320 \, \cancel{K})}{293 \, \cancel{K}} = 2.51 \text{ atm of air in the tire}$

Chapter 5 Gases, Liquids, and Solids

5.17 Charles's Law:

at constant presure, $\left(\dfrac{\cancel{P_1}V_1}{T_1}\right) = \left(\dfrac{\cancel{P_2}V_2}{T_2}\right)$ reduces to $\dfrac{V_1}{T_1} = \dfrac{V_2}{T_2}$

$$V_2 = \frac{V_1 T_2}{T_1} = \frac{(4.17\ L)(448\ \cancel{K})}{998\ \cancel{K}} = 1.87\ L \text{ of ethane gas upon cooling}$$

5.19 Gay-Lussac's Law:

at constant volume, $\left(\dfrac{P_1\cancel{V_1}}{T_1}\right) = \left(\dfrac{P_2\cancel{V_2}}{T_2}\right)$ reduces to $\dfrac{P_1}{T_1} = \dfrac{P_2}{T_2}$

$$T_2 = \frac{P_2 T_1}{P_1} = \frac{(375\ mm\ Hg)(898\ K)}{450\ mm\ Hg} = 748\ K\ (475°C)$$

5.21 Gay-Lussac's Law:

at constant volume, $\left(\dfrac{P_1\cancel{V_1}}{T_1}\right) = \left(\dfrac{P_2\cancel{V_2}}{T_2}\right)$ reduces to $\dfrac{P_1}{T_1} = \dfrac{P_2}{T_2}$

$$P_2 = \frac{P_1 T_2}{T_1} = \frac{(1.00\ atm)(438\ \cancel{K})}{373\ \cancel{K}} = 1.17\ atm$$

5.23 Complete this table: Use the $\dfrac{P_1 V_1}{T_1} = \dfrac{P_2 V_2}{T_2}$ equation.

V₁	T₁	P₁	V₂	T₂	P₂
546 L	43°C	6.5 atm	**1198 L**	65°C	1.9 atm
43 mL	-56°C	865 torr	**48 mL**	43°C	1.50 atm
4.2 L	234 K	0.87 atm	3.2 L	29°C	**1.5 atm**
1.3 L	25°C	740 mm Hg	**1.2 L**	0°C	1.0 atm

5.25 Charles's Law: atmospheric pressure acting on balloon is constant

$$V_2 = \frac{V_1 T_2}{T_1} = \frac{(1.2\ L)(77\ \cancel{K})}{298\ \cancel{K}} = 0.31\ L \text{ balloon's final volume}$$

5.27 $P_2 = \dfrac{P_1 V_1 T_2}{T_1 V_2} = \dfrac{(56.44\ \cancel{L})(2.00\ atm)(281\ \cancel{K})}{(310\ \cancel{K})(23.52\ \cancel{L})} = 4.35\ atm$

5.29 $V_2 = \dfrac{P_1 V_1 T_2}{T_1 P_2} = \dfrac{(756 \ \text{mm Hg})(30.0 \ \text{mL})(260.5 \ K)}{(298 \ K)(325 \ \text{mm Hg})} = 61.0 \ \text{mL}$

5.31 (a) $n = \dfrac{PV}{RT} = \dfrac{(1.33 \ \text{atm})(50.3 \ L)}{(0.0821 \ L \cdot \text{atm} \cdot \text{mol}^{-1} \cdot K^{-1})(350 \ K)} = 2.33 \ \text{mol}$

(b) The only information that we need to know about the gas is that it is an ideal gas.

5.33 Using the PV=nRT Gas Law equation, the following equation is derived:

$$MW = \dfrac{(\text{mass})RT}{PV} = \dfrac{(8.00 \ \text{g})(0.0821 \ L \cdot \text{atm} \cdot \text{mol}^{-1} \cdot K^{-1})(273 \ K)}{(2.00 \ \text{atm})(22.4 \ L)} = 4.00 \ \text{g/mol}$$

5.35 From the PV=nRT equation, we can derive:

$$\dfrac{\text{mass}}{V} = \text{density} = \dfrac{P(MW)}{RT}$$

(a) At constant T, equation reduces to density = (constant) x (pressure), therefore, the density increases as pressure increases.

(b) At constant P, the equation reduces to density = (constant)(1/T), therefore, density decreases with increasing T.

5.37 From the PV = nRT equation:

(a) $n = \dfrac{PV}{RT} = \dfrac{(3.00 \ \text{atm})(200 \ L)}{(0.0821 \ L \cdot \text{atm} \cdot \text{mol}^{-1} \cdot K^{-1})(296 \ K)} = 24.7 \ \text{mol} \ O_2$

(b) $\text{mass of } O_2 = 24.7 \ \text{mol} \ O_2 \left(\dfrac{32.0 \ \text{g} \ O_2}{1 \ \text{mol} \ O_2} \right) = 790 \ \text{g} \ O_2$

5.39 $5.5 \ L \ \text{air} \left(\dfrac{0.21 \ L \ O_2}{1 \ L \ \text{air}} \right) = 1.16 \ L \ O_2$

$\text{moles of } O_2 = n = \dfrac{PV}{RT} = \dfrac{(1.1 \ \text{atm})(5.5 \ L)}{(0.0821 \ L \cdot \text{atm} \cdot \text{mol}^{-1} \cdot K^{-1})(310 \ K)} = 0.238 \ \text{mol} \ O_2$

$0.238 \ \text{mol} \ O_2 \left(\dfrac{6.02 \times 10^{23} \ \text{molecules} \ O_2}{1 \ \text{mol} \ O_2} \right) = 1.4 \times 10^{23} \ \text{molecules} \ O_2$

5.41 1.0000 mole air = 0.7808 mol N_2 + 0.2095 mol O_2 + 0.0093 mol Ar

MW_{air} =

$$0.7808 \text{ mol}_{N_2}\left(\frac{28.01 \text{ g } N_2}{1 \text{ mol } N_2}\right) + 0.2095 \text{ mol}_{O_2}\left(\frac{32.00 \text{ g } O_2}{1 \text{ mol } O_2}\right) + 0.0093 \text{ mol}_{Ar}\left(\frac{39.95 \text{ g Ar}}{1 \text{ mol Ar}}\right)$$

(a) MW_{air} = 28.95 g/mol

(b) $d(g/L)_{air} = \left(\frac{28.95 \text{ g air}}{1 \text{ mol air}}\right)\left(\frac{1 \text{ mol air}}{22.4 \text{ L}}\right) = 1.29$ g/L

5.43 Molecules in the gas phase are in constant, random motion. Over time, the gas molecules mix thoroughly with air and, and in so doing, travel to the stratosphere.

5.45 Ideal Gas Law:

$$100 \text{ L CO}\left(\frac{1 \text{ mol CO}}{22.4 \text{ L CO}}\right)\left(\frac{6.02 \times 10^{23} \text{ molecules}}{1 \text{ mol CO}}\right) = 2.69 \times 10^{24} \text{ molcules of CO}$$

5.47 From the equation: $2NaN_3(s) \rightarrow 2Na(s) + 3N_2(g)$

$$100 \text{ g NaN}_3\left(\frac{1 \text{ mol NaN}_3}{65.01 \text{ g NaN}_3}\right)\left(\frac{3 \text{ mol } N_2}{2 \text{ mol NaN}_3}\right) = 2.31 \text{ mol } N_2$$

$$V_{N_2} = \frac{nRT}{P} = \frac{(2.31 \text{ mol } N_2)(0.0821 \text{ L} \cdot \text{atm} \cdot \text{mol}^{-1} \cdot K^{-1})(300 \text{ K})}{1.00 \text{ atm}} = 56.9 \text{ L}$$

5.49 $P_T = P_{N2} + P_{O2} + P_{CO2} + P_{H2O}$

P_{N_2} = (0.740)(1.0 atm) = 0.740 atm (562.4 mm Hg)

P_{O_2} = (0.194)(1.0 atm) = 0.194 atm (147.5 mm Hg)

P_{H_2O} = (0.062)(1.0 atm) = 0.062 atm (47.1 mm Hg)

P_{CO_2} = (0.004)(1.0 atm) = 0.004 atm (3.0 mm Hg)

P_T = 1.0 atm   (760.0 mm Hg)

5.51 Covalent bonds are stronger than H-bonds. Covalent bonds involve the sharing of electrons, where H-bonds involve weaker electrostatic interactions.

5.53 Yes, the water OH can H-bond (H-bond donor) with the oxygen lone pair on the S=O (H-bond acceptor).

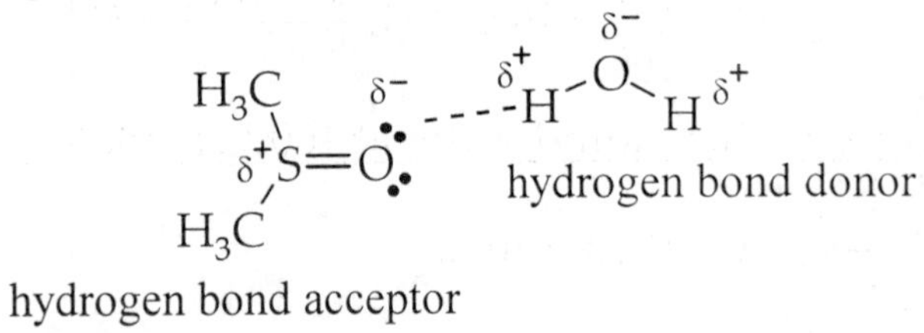

hydrogen bond acceptor

5.55 Ethanol is a polar molecule and engages in intermolecular hydrogen bonding. Carbon dioxide is a nonpolar molecule and has only weak London Dispersion intermolecular forces. The stronger hydrogen bonding intermolecular forces require more energy and higher temperatures to break before boiling.

5.57 Hexane has a higher boiling point. It is a larger molecule than butane, thus hexane has larger London Dispersion forces to overcome before boiling.

5.59 $SH = \dfrac{Heat}{m(T_2 - T_1)} = \dfrac{170 \text{ cal}}{(36.6 \text{ g})(50^\circ C - 30^\circ C)} = 0.232 \text{ cal} \cdot g^{-1} \cdot {}^\circ C^{-1}$

5.61 heat = mass × SH × $(T_2 - T_1)$ = (200.59 g Hg)(0.0332 cal·g^{-1}·$^\circ C^{-1}$)(36 $^\circ C$) = 240 cal

5.63 CH_4 has a much higher vapor pressure than water at room temperature because the boiling point of CH_4 has already been exceeded (vapor pressure equal atmospheric pressure. At room temperature, water has not reached its boiling point (100°C) thus its vapor pressure is well below atmospheric pressure.

5.65 Energy required to heat 1 mol of ice (0°C to a liquid at 23°C) = heat + heat of fusion

$$energy_{(total)} = 18.02 \text{ g } H_2O \left(\frac{1 \text{ cal}}{g \cdot {}^\circ C} \right) (23^\circ C) + 18.02 \text{ g } H_2O \left(\frac{80 \text{ cal}}{g} \right) = 1860 \text{ cal}$$

5.67 (a) The energy released when steam at 100°C is condensed to 37°C:

$$10 \; \cancel{g} \; H_2O \left(\frac{1.0 \; cal}{\cancel{g} \cdot {}^\circ \cancel{C}} \right) \left(63^\circ \cancel{C} \right) + 10 \; g \; H_2O \left(\frac{540 \; cal}{1g} \right) = 6030 \; cal \; (6.0 \; kcal)$$

(b) The energy released when liquid water at 100°C is cooled to 37°C:

$$10 \; \cancel{g} \; H_2O \left(\frac{1.0 \; cal}{\cancel{g} \cdot {}^\circ \cancel{C}} \right) \left(63^\circ \cancel{C} \right) = 630 \; cal$$

(c) Steam releases its heat of vaporization when condensed from gas to liquid. The heat of vaporization of liquid water (540 cal/g) is much greater than the heat capacity of water (1 cal/g · °C). More heat is released to the skin by condensing steam than by liquid water cooling from 100°C to 20°C.

5.69 The dry ice sublimes, forming a gas. The Ideal Gas Law equation becomes:

$$V = \frac{nRT}{P} = \frac{\left(\dfrac{156 \; \cancel{g} \; CO_2}{44.01 \; \cancel{g}/mol \; CO_2} \right) \left(0.0821 \; L \cdot \cancel{atm} \cdot \cancel{mol^{-1}} \cdot \cancel{K^{-1}} \right) \left(298 \; \cancel{K} \right)}{\left(740 \; \cancel{mm \; Hg} \right) \left(\dfrac{1 \; atm}{760 \; \cancel{mm \; Hg}} \right)} = 89.1 \; L$$

5.71 A phase diagram of water shows that heating ice from -10°C to 20°C while reducing the pressure from 1 atm to 0.1 atm results in sublimation.

5.73 Soot is a form of carbon that has the highest entropy (greatest disorder). The carbons in soot are randomly arranged. Diamond and graphite are crystalline solids where carbon atoms have ordered arrangements in a lattice. Buckyballs have carbon atoms ordered within a molecule. If you look closely at newly discovered Buckyball, you will see that it is the only pure form of carbon. The other three forms all have hydrogen atoms terminating the carbons on the edges of the molecules!

5.75 Carbon monoxide is bound to hemoglobin and does not allow it to carry oxygen. Oxygen under pressure in a hyperbaric chamber dissolves in the plasma and is carried to the tissues without the aid of hemoglobin.

5.77 Frostbite occurs when tissue freezes. The frozen water in the cells expands and ruptures the cells, causing damage, sometimes irreversible.

5.79 Supercritical carbon dioxide has the density of a liquid but maintains its gas like property of being able to flow with little viscosity or surface tension.

Chapter 5 Gases, Liquids, and Solids

5.81 (a) Gas pressure results from the forces of gas molecule collisions with the container walls that hold the gas.
(b) Temperature is a measure of the average kinetic energy of the gas molecules.

5.83 Gay-Lussac's Law says that the pressure inside the can will more than double.

$$P_2 = \frac{P_1 T_2}{T_1} = \frac{(3.00 \text{ atm})(673 \text{ K})}{296 \text{ K}} = 6.82 \text{ atm}$$

5.85 The barometric pressure drops when the air becomes less dense.

5.87 Carbon monoxide (CO) has a dipole moment, therefore has greater intermolecular interaction through dipole-dipole attractions. Carbon dioxide has no dipole moment; therefore it only has weak intermolecular interactions through London dispersion forces.

5.89 The following conversion must be calculated first:

$$P \text{ (in atm)} = 35 \text{ inch Hg} \left(\frac{25.4 \text{ mm Hg}}{1 \text{ inch Hg}} \right) \left(\frac{1 \text{ atm}}{760 \text{ mm Hg}} \right) = 1.17 \text{ atm}$$

$$n = \frac{PV}{RT} = \frac{(1.17 \text{ atm})(10 \text{ L})}{(0.0821 \text{ L} \cdot \text{atm} \cdot \text{mol}^{-1} \cdot \text{K}^{-1})(300 \text{ K})} = 0.48 \text{ mol N}_2$$

$$n_2 = \frac{P_2 n_1}{P_1} = \frac{(60 \text{ in Hg})(0.48 \text{ mol})}{35 \text{ in Hg}} = 0.82 \text{ mol N}_2$$

0.82 mol - 0.48 mol = 0.34 mol N_2 must be added to the flask

5.91 Gases are transparent because of the large amount of empty space between gas molecules. Visible light passes through a gas sample without much interaction with the gas molecules.

5.93 Pentane has the lower boiling point, therefore it would be predicted to have the higher vapor pressure at 20°C.

5.95 Water has hydrogen bonding that holds it in the liquid state, giving it a higher boiling point than hydrogen sulfide, which does not engage in hydrogen bonding.

5.97 Balanced equation: $4Al(s) + 3O_2(g) \rightarrow 2Al_2O_3(s)$

moles of O_2 needed = 3.42 g Al $\left(\dfrac{1 \text{ mol Al}}{26.982 \text{ g Al}} \right) \left(\dfrac{3 \text{ mol } O_2}{4 \text{ mol Al}} \right)$ = 0.0951 mol O_2

mol of air needed = 0.0951 mol O_2 $\left(\dfrac{1 \text{ mol air}}{0.21 \text{ mol } O_2} \right)$ = 0.453 mol air

$V_{air} = \dfrac{nRT}{P} = \dfrac{(0.453 \text{ mol})(0.0821 \text{ L} \cdot \text{atm} \cdot \text{mol}^{-1} \cdot K^{-1})(298 \ K)}{(0.975 \text{ atm})}$ = 11.4 L air

5.99 The phase diagram indicates that water will be in the vapor phase at 0.2 atm and 120°C.

Chapter 6 Solutions and Colloids

<u>6.1</u> 4.4% w/v KBr solution = 4.4 g KBr in 100 mL of solution

$$250 \text{ mL solution} \left(\frac{4.4 \text{ g KBr}}{100 \text{ mL solution}} \right) = 11 \text{ g KBr}$$

Add enough water to 11 g KBr to make 250 mL of solution

<u>6.3</u> First, calculate the number of moles and mass of KCl that are needed: moles = M x V.

$$\text{mol KCl} = \left(\frac{1.06 \text{ mol KCl}}{1 \text{ L sol}} \right) (2.0 \text{ L sol}) = 2.12 \text{ mol KCl}$$

$$\text{mass of KCl} = 2.12 \text{ mol KCl} \left(\frac{74.6 \text{ g KCl}}{1 \text{ mol KCl}} \right) = 158 \text{ g KCl}$$

Place 158 g of KCl into a 2-L volumetric flask, add some water, swirl until the solid has dissolved, and then fill the flask with water to the 2.0-L mark.

<u>6.5</u> First, convert grams of glucose into moles of glucose, then convert moles of glucose into mL of solution:

$$\text{mol of glucose} = 10.0 \text{ g glucose} \left(\frac{1 \text{ mol glucose}}{180 \text{ g glucose}} \right) = 0.0556 \text{ mol glucose}$$

$$0.0556 \text{ mol glucose} \left(\frac{1 \text{ L sol}}{0.300 \text{ mol glucose}} \right) \left(\frac{1000 \text{ mL sol}}{1 \text{ L sol}} \right) = 185 \text{ mL glucose sol}$$

<u>6.7</u> Use the $M_1 V_1 = M_2 V_2$ equation:

$$V_1 = \frac{(0.600 \text{ } M \text{ HCl})(300 \text{ mL sol HCl})}{(12.0 \text{ } M \text{ HCl})} = 15.0 \text{ mL}$$

Place 15.0 mL of a 12.0 M HCl solution into a 300-mL volumetric flask, add some water, swirl until completely mixed, and then fill the flask with water to the 300-mL mark.

6.9 $\Delta T = (1.86°C/mol)$(mole of particles in solution per 1000 g of water):

$$\text{mol CH}_3\text{OH} = 215 \text{ g CH}_3\text{OH} \left(\frac{1 \text{ mol CH}_3\text{OH}}{32.0 \text{ g CH}_3\text{OH}} \right) = 6.72 \text{ mol CH}_3\text{OH}$$

$\Delta T = (1.86°C/mol)(6.72 \text{ mol}) = 12.5°C$ freezing point will be lowered by 12.5°C

new freezing point$_{(H_2O)}$ = 0°C - 12.5°C = -12.5°C

6.11 Osmolarity = $M \times i$
First, calculate the molarity (M):

$$M_{\text{Na}_3\text{PO}_4} = \frac{3.3 \text{ g Na}_3\text{PO}_4}{100 \text{ mL sol}} \left(\frac{1 \text{ mol Na}_3\text{PO}_4}{163.9 \text{ g Na}_3\text{PO}_4} \right) \left(\frac{1000 \text{ mL sol}}{1 \text{ L sol}} \right) = 0.20 \text{ } M \text{ Na}_3\text{PO}_4$$

A Na_3PO_4 molecule gives 3 Na^+ ions and 1 PO_4^{3-} ion for a total of 4 solute particles.
Osmolarity = $(0.20 \text{ } M)(4 \text{ ions}) = 0.80$ osmol

6.13 Vinegar is an aqueous solution of acetic acid, therefore water is the solvent.

6.15 (a) both tin and copper are solids
      (b) solid solute (caffeine, flavorings) and liquid solvent (water)
      (c) both CO_2 and steam (H_2O) are gases
      (d) gas (CO_2) and liquid (ethyl alcohol) solutes in liquid (water) solvent

6.17 Mixtures of gasses are true solutions because they mix in all proportions, molecules are distributed uniformly, and the component gasses do not separate upon standing.

6.19 You get an unsaturated solution. A 50 mL solution of this compound becomes saturated with 1.25 g solute.

6.21 (c) $CH_3CH_2CH_2CH_3$ (butane) is the least polar of the choices and would be the most soluble in a non-polar benzene.

6.23 Remember, compounds of similar polarity are miscible.
      (a) H_2O and CH_3OH will be miscible because both are polar and can H-bond.
      (b) H_2O (polar) and C_6H_6 (non-polar) will not be miscible.
      (c) C_6H_{14} and CCl_4 will be miscible because both are non-polar.
      (d) CCl_4 (non-polar) and CH_3OH (polar) will not be miscible.

Chapter 6 Solutions and Colloids

6.25 The solubility of oxygen in water decreases as the temperature of water increases. The concentration of oxygen in the warm water may become low enough where there is not enough for fish to survive.

6.27 According to Henry's Law, the solubility of a gas in a liquid is directly proportional to the pressure; therefore, the solubility of ammonia in water will be (a) greater at 2 atm compared to 0.5 atm of pressure.

6.29 (a) vol of ethanol = 280 mL sol $\left(\dfrac{27 \text{ mL ethanol}}{100 \text{ mL sol}}\right)$ = 76 mL ethanol

76 mL ethanol dissolved in 204 mL water (to give 280 mL of solution)

(b) vol of ethyl acetate = 435 mL sol $\left(\dfrac{1.8 \text{ mL ethyl acetate}}{100 \text{ mL sol}}\right)$ = 7.8 mL ethyl acetate

8 mL ethyl acetate dissolved in 427 mL water (to give 435 mL of solution)

(c) vol of benzene = 1.65 L sol $\left(\dfrac{1000 \text{ mL sol}}{1 \text{ L sol}}\right)\left(\dfrac{8.00 \text{ mL benzene}}{100 \text{ mL sol}}\right)$ = 132 mL benzene

0.13 L benzene dissolved in 1.52 L chloroform (to give 1.65 L of solution)

6.31 (a) % (w/v) = $\dfrac{623 \text{ mg casein}}{15.0 \text{ mL sol}}\left(\dfrac{1 \text{ g casein}}{1000 \text{ mg casein}}\right)$ × 100% = 4.15 % w/v casein

(b) % (w/v) = $\dfrac{74 \text{ mg vit. C}}{250 \text{ mL sol}}\left(\dfrac{1 \text{ g vit. C}}{1000 \text{ mg vit. C}}\right)$ × 100% = 0.030 % w/v vitamin C

(c) % (w/v) = $\dfrac{3.25 \text{ g sucrose}}{186 \text{ mL sol}}$ × 100% = 1.75 % w/v sucrose

<u>6.33</u> (a) $175 \text{ mL sol} \left(\dfrac{1 \text{ L sol}}{1000 \text{ mL sol}} \right) \left(\dfrac{1.14 \text{ mol NH}_4\text{Br}}{1 \text{ L solution}} \right) \left(\dfrac{97.9 \text{ g NH}_4\text{Br}}{1 \text{ mol NH}_4\text{Br}} \right) = 19.5 \text{ g NH}_4\text{Br}$

Place 19.5 g of NH₄Br into a 175-mL volumetric flask, add some water, swirl until completely dissolved, and then fill the flask with water to the 175-mL mark.

(b) $1.35 \text{ L sol} \left(\dfrac{0.825 \text{ mol NaI}}{1 \text{ L solution}} \right) \left(\dfrac{149.9 \text{ g NaI}}{1 \text{ mol NaI}} \right) = 167 \text{ g NaI}$

Place 167 g of NaI into a 1.35-L volumetric flask, add some water, swirl until completely dissolved, and then fill the flask with water to the 1.35-L mark.

(c) $330 \text{ mL sol} \left(\dfrac{1 \text{ L sol}}{1000 \text{ mL sol}} \right) \left(\dfrac{0.16 \text{ mol ethanol}}{1 \text{ L solution}} \right) \left(\dfrac{46.1 \text{ g ethanol}}{1 \text{ mol ethanol}} \right) = 2.4 \text{ g ethanol}$

Place 2.4 g of ethanol into a 330-mL volumetric flask, add some water, swirl until completely mixed, and then fill the flask with water to the 330-mL mark.

<u>6.35</u> $M_{NaCl} = \dfrac{5.0 \text{ mg NaCl}}{0.5 \text{ mL sol}} \left(\dfrac{1 \text{ g NaCl}}{1000 \text{ mg NaCl}} \right) \left(\dfrac{1 \text{ mol NaCl}}{58.4 \text{ g NaCl}} \right) \left(\dfrac{1000 \text{ mL sol}}{1 \text{ L sol}} \right) = 0.2 \, M \text{ NaCl}$

<u>6.37</u> $M_{glucose} = \dfrac{22.0 \text{ g glucose}}{240 \text{ mL sol}} \left(\dfrac{1 \text{ mol glucose}}{180 \text{ g glucose}} \right) \left(\dfrac{1000 \text{ mL sol}}{1 \text{ L sol}} \right) = 0.509 \, M \text{ glucose}$

$M_{K+} = \dfrac{190 \text{ mg K}^+}{240 \text{ mL sol}} \left(\dfrac{1 \text{ g}}{1000 \text{ mg}} \right) \left(\dfrac{1 \text{ mol K}^+}{39.1 \text{ g K}^+} \right) \left(\dfrac{1000 \text{ mL sol}}{1 \text{ L sol}} \right) = 0.0202 \, M \text{ K}^+$

$M_{Na+} = \dfrac{4.00 \text{ mg K}^+}{240 \text{ mL sol}} \left(\dfrac{1 \text{ g}}{1000 \text{ mg}} \right) \left(\dfrac{1 \text{ mol Na}^+}{23.0 \text{ g Na}^+} \right) \left(\dfrac{1000 \text{ mL sol}}{1 \text{ L sol}} \right) = 7.25 \times 10^{-4} \, M \text{ Na}^+$

<u>6.39</u> $M_{sucrose} = \dfrac{13 \text{ g sucrose}}{15 \text{ mL sol}} \left(\dfrac{1 \text{ mol sucrose}}{342.3 \text{ g sucrose}} \right) \left(\dfrac{1000 \text{ mL sol}}{1 \text{ L sol}} \right) = 2.5 \, M \text{ sucrose}$

<u>6.41</u> Use the $\%_1 V_1 = \%_2 V_2$ equation:

$V_2 = \dfrac{(0.750\% \text{ w/v albumin})(5.00 \text{ mL sol})}{(0.125\% \text{ w/v albumin})} = 30.0 \text{ mL}$

The total volume of the dilution is 30.0 mL. Starting with a 5.00 mL solution, <u>25.0 mL of water must be added</u> to reach a final volume of 30.0 mL.

<u>6.43</u> Use the $\%_1V_1 = \%_2V_2$ equation:

$$V_1 = \frac{(0.25\% \text{ w/v } H_2O_2)(250 \text{ mL sol})}{(30.0\% \text{ w/v } H_2O_2)} = 2.1 \text{ mL } H_2O_2$$

Place 2.1 mL of 30.0% w/v H_2O_2 into a 250-mL volumetric flask, add some water, swirl until completely mixed, and then fill the flask with water to the 250-mL mark.

<u>6.45</u> (a) $\dfrac{12.5 \text{ mg Captopril}}{325 \text{ mg pill}} \times 10^6 = 3.85 \times 10^4 \text{ ppm Captopril}$

(b) $\dfrac{22 \text{ mg Mg}^{2+}}{325 \text{ mg pill}} \times 10^6 = 6.8 \times 10^4 \text{ ppm Mg}^{2+}$

(c) $\dfrac{0.27 \text{ mg Ca}^{2+}}{325 \text{ mg pill}} \times 10^6 = 8.3 \times 10^2 \text{ ppm Ca}^{2+}$

<u>6.47</u> Assume the density of the lake water to be 1.0 g/mL

$$1 \times 10^7 \text{ L water} \left(\frac{1000 \text{ mL water}}{1 \text{ L water}} \right) \left(\frac{1.0 \text{ g water}}{1.0 \text{ mL water}} \right) = 1 \times 10^{10} \text{ g water}$$

$$\frac{0.1 \text{ g dioxin}}{1 \times 10^{10} \text{ g water}} \times 10^9 = 0.01 \text{ ppb dioxin}$$

No, the dioxin level in the lake did not reach a dangerous level.

<u>6.49</u> First, calculate the mass (in grams) of the nutrients in the cheese based on their percentages of daily allowances, then calculate concentration in ppm:

Iron: $(0.02)(15 \text{ mg Fe}) \left(\dfrac{1 \text{ g Fe}}{1000 \text{ mg Fe}} \right) = 3 \times 10^{-4} \text{ g Fe}$

$$\frac{3 \times 10^{-4} \text{ g Fe}}{28 \text{ g cheese}} \times 10^6 = 10 \text{ ppm Fe}$$

Calcium: $(0.06)(1200 \text{ mg Ca}) \left(\dfrac{1 \text{ g Ca}}{1000 \text{ mg Ca}} \right) = 7 \times 10^{-2} \text{ g Ca}$

$$\frac{7 \times 10^{-2} \text{ g Ca}}{28 \text{ g cheese}} \times 10^6 = 3 \times 10^3 \text{ ppm Ca}$$

Vitamin A: $(0.06)(0.800 \text{ mg Vit. A}) \left(\dfrac{1 \text{ g}}{1000 \text{ mg Vit. A}} \right) = 5 \times 10^{-5} \text{ g Vitamin A}$

$$\frac{5 \times 10^{-5} \text{ g Vit. A}}{28 \text{ g cheese}} \times 10^6 = 2 \text{ ppm Vitamin A}$$

6.51 Both (a) 0.1 M KCl and (b) 0.1 M $(NH_4)_3PO_4$ will conduct electricity because they are strong electrolytes. A 0.1 M $(NH_4)_3PO_4$ solution (b) will have a greater conductivity than 0.1 M KCl because it has the largest concentration of dissociated ions (4 mole of ions dissociated per mole of $(NH_4)_3PO_4$, giving a 0.4 M solution of ions) compared to a 0.1 M KCl solution (2 mole of ions dissociated per mole of KCl, giving a 0.2 M solution of ions).

6.53 The polar compounds, (b), (c), and (d) will be soluble in water by forming hydrogen bonds with water.

6.55 (a) homogeneous   (b) heterogeneous   (c) colloid
      (d) heterogeneous   (e) colloid   (f) colloid

6.57 As the temperature of the solution decreased, the protein molecules must have aggregated and formed a colloidal mixture. The turbid appearance is the result of the Tyndall effect.

6.59 ΔT = (1.86°C/mol)(mole of particles in solution per 1000 g of water):

(a) $\Delta T = 1 \text{ mol NaCl} \left(\dfrac{2 \text{ mol particle}}{1 \text{ mol NaCl}} \right) \left(\dfrac{1.86°C}{\text{mol particle}} \right) = 3.72°C$ f.p. = -3.72°C

(b) $\Delta T = 1 \text{ mol MgCl}_2 \left(\dfrac{3 \text{ mol particle}}{1 \text{ mol MgCl}_2} \right) \left(\dfrac{1.86°C}{\text{mol particle}} \right) = 5.58°C$ f.p. = -5.58°C

(c) $\Delta T = 1 \text{ mol (NH}_4)_2CO_3 \left(\dfrac{3 \text{ mol particle}}{1 \text{ mol (NH}_4)_2CO_3} \right) \left(\dfrac{1.86°C}{\text{mol particle}} \right) = 5.58°C$

      f.p. = -5.58°C

(d) $\Delta T = 1 \text{ mol Al(HCO}_3)_3 \left(\dfrac{4 \text{ mol particle}}{1 \text{ mol Al(HCO}_3)_3} \right) \left(\dfrac{1.86°C}{\text{mol particle}} \right) = 7.44°C$

      f.p. = -7.44°C

6.61 Methanol is a non-electrolyte and does not dissociate. Use the following equation for freezing point depression while also converting to grams.

$$\text{moles of CH}_3\text{OH per 1000 g H}_2\text{O} = \Delta T \left(\dfrac{1 \text{ mol particles}}{1.86°C} \right)$$

$$20°C \left(\dfrac{1 \text{ mol particles}}{1.86°C} \right) \left(\dfrac{1 \text{ mole CH}_3\text{OH}}{1 \text{ mol particles}} \right) \left(\dfrac{32.0 \text{ g CH}_3\text{OH}}{1 \text{ mol CH}_3\text{OH}} \right) = 344 \text{ g CH}_3\text{OH}$$

6.63 Acetic acid is a weak acid, therefore does not completely dissociate into ions. KF is a strong electrolyte, completely dissociating into two ions and doubling the effect on freezing point depression compared to acetic acid.

6.65 In each case, the side with the greater osmolarity rises.
   (a) B   (b) B   (c) A   (d) B   (e) A   (f) same

6.67 (a) $osmol_{Na_2CO_3} = 0.39\ M \times 3$ particles = 1.2 osmol

   (b) $osmol_{Al(NO_3)_3} = 0.62\ M \times 4$ particles = 2.5 osmol

   (c) $osmol_{LiBr} = 4.2\ M \times 2$ particles = 8.4 osmol

   (d) $osmol_{K_3PO_4} = 0.009\ M \times 4$ particles = 0.04 osmol

6.69 Cells in hypertonic solutions under crenation (shrink).

$$osmol_{NaCl} = \frac{0.9\ g\ NaCl}{100\ mL\ sol}\left(\frac{1000\ mL\ sol}{1\ L\ sol}\right)\left(\frac{1\ mol\ NaCl}{58.4\ g\ NaCl}\right)\left(\frac{2\ mol\ particles}{1\ mol\ NaCl}\right) = 0.3\ osmol$$

(a) $$\frac{0.3\ g\ NaCl}{100\ mL\ sol}\left(\frac{1000\ mL\ sol}{1\ L\ sol}\right)\left(\frac{1\ mol\ NaCl}{58.4\ g\ NaCl}\right)\left(\frac{2\ mol\ particles}{1\ mol\ NaCl}\right) = 0.1\ osmol\ NaCl$$

(b) $osmol_{glucose} = 0.9\ M \times 1$ particle = 0.9 osmol

(c) $$\frac{0.9\ g\ glu}{100\ mL\ sol}\left(\frac{1000\ mL\ sol}{1\ L\ sol}\right)\left(\frac{1\ mol\ glu}{180\ g\ glu}\right)\left(\frac{1\ mol\ particles}{1\ mol\ glu}\right) = 0.05\ osmol\ glucose$$

   Solution (b) has a concentration greater than the isotonic solution so it will crenate red blood cells.

6.71 Carbon dioxide (CO_2) dissolves in rain water to form a dilute solution of carbonic acid (H_2CO_3), which is a weak acid.

6.73 Nitrogen dissolved in the blood can lead to a narcotic effect referred to as "rapture of the deep", which is similar to alcohol-induced intoxication.

6.75 The main component of limestone and marble is calcium carbonate ($CaCO_3$).

6.77 The minimum pressure required for the reverse osmosis in the desalinization of sea water exceeds 100 atm (the osmotic pressure of sea water).

Chapter 6 Solutions and Colloids

6.79 $$\frac{0.2 \text{ g NaHCO}_3}{100 \text{ mL sol}}\left(\frac{1000 \text{ mL sol}}{1 \text{ L sol}}\right)\left(\frac{1 \text{ mol NaHCO}_3}{84.0 \text{ g NaHCO}_3}\right)\left(\frac{2 \text{ mol particles}}{1 \text{ mol NaHCO}_3}\right) = 0.05 \text{ osmol}$$

$osmol_{NaHCO_3} = 0.05$ osmol

$$\frac{0.2 \text{ g KHCO}_3}{100 \text{ mL sol}}\left(\frac{1000 \text{ mL sol}}{1 \text{ L sol}}\right)\left(\frac{1 \text{ mol KHCO}_3}{100.1 \text{ g KHCO}_3}\right)\left(\frac{2 \text{ mol particles}}{1 \text{ mol KHCO}_3}\right) = 0.04 \text{ osmol}$$

$osmol_{KHCO_3} = 0.04$ osmol

Yes, the change made a change in the tonicity. The error in replacing NaHCO₃ with KHCO₃ resulted in a hypotonic solution and an electrolyte imbalance by reducing the number of ions (osmolarity) in solution.

6.81 The ethanol displaced the water from the solvation layer of the hyaluronic acid and thus allowed hyaluronic acid molecules to stick together upon collision and aggregate.

6.83 $$\frac{2 \times 10^{-5} \text{ mol As}_2\text{O}_3}{1 \text{ L sol}}\left(\frac{198 \text{ g As}_2\text{O}_3}{1 \text{ mol As}_2\text{O}_3}\right)\left(\frac{1 \text{ L sol}}{1000 \text{ mL sol}}\right)\left(\frac{1 \text{ mL sol}}{1 \text{ g sol}}\right) \times 10^6 = 4 \text{ ppm As}_2\text{O}_3$$

6.85 Methanol is more efficient at lowering the freezing point of water. It has a lower molecular weight; hence there are more moles of it compared to an equal mass of antifreeze.

6.87 $CO_2(g) + H_2O(l) \rightarrow H_2CO_3(aq)$ carbonic acid
$SO_2(g) + H_2O(g) \rightarrow H_2SO_3(aq)$ sulfurous acid

6.89 Use the $\%_1V_1 = \%_2V_2$ equation:
$$V_1 = \frac{(4.5\% \text{ w/v HNO}_3)(300 \text{ mL sol})}{(35\% \text{ w/v HNO}_3)} = 39 \text{ mL HNO}_3$$
Place 39 mL of 35% w/v HNO₃ into a 300-mL volumetric flask, add some water, swirl until completely mixed, and then fill the flask with water to the 300-mL mark.

6.91 $$1016 \text{ kg H}_2\text{O}\left(\frac{1000 \text{ g H}_2\text{O}}{1 \text{ kg H}_2\text{O}}\right)\left(\frac{6 \text{ g pollutant}}{10^9 \text{ g H}_2\text{O}}\right) = 6 \times 10^{-3} \text{ g pollutant}$$

Chapter 6 Solutions and Colloids

<u>6.93</u> Assume that the density of the pool water is 1.00 g/mL

$$Cl_2 \text{ (ppm)} = \frac{0.00500 \text{ mol } Cl_2}{1 \text{ L sol}}\left(\frac{70.9 \text{ g } Cl_2}{1 \text{ mol } Cl_2}\right)\left(\frac{1 \text{ L sol}}{1000 \text{ mL sol}}\right)\left(\frac{1 \text{ mL sol}}{1 \text{ g sol}}\right) \times 10^6 = 355 \text{ ppm}$$

$$20,000 \text{ L Pool } H_2O\left(\frac{0.00500 \text{ mol } Cl_2}{1 \text{ L Pool } H_2O}\right)\left(\frac{70.9 \text{ g } Cl_2}{1 \text{ mol } Cl_2}\right)\left(\frac{1 \text{ kg } Cl_2}{1000 \text{ g } Cl_2}\right) = 7.09 \text{ kg } Cl_2 \text{ added}$$

Chapter 7 Reaction Rates and Chemical Equilibrium

7.1 The rate of the reaction is equal to the change in the amount of O_2 per unit time.

$$\text{rate of } O_2 \text{ formation} = \frac{(0.35 \text{ L } O_2 - 0.020 \text{ L } O_2)}{15 \text{ min}} = 0.022 \text{ L } O_2/\text{min}$$

7.3 $\quad K = \dfrac{[H_2SO_4]}{[SO_3][H_2O]}$

7.5 $\quad K = \dfrac{[PCl_5]}{[PCl_3][Cl_2]} = \dfrac{[1.66 \, M]}{[1.66 \, M][1.66 \, M]} = 0.602 \, M^{-1}$

7.7 Le Chatelier's principle would predict that the equilibrium would shift to the left by adding a product (Br_2).

7.9 If the equilibrium shifts right with the addition of heat, heat must have been a reactant and the reaction endothermic.

7.11 $\quad \text{rate of } CH_3I \text{ formation} = \dfrac{(0.840 \, M \, CH_3I - 0.260 \, M \, CH_3I)}{80 \text{ min}} = 7.25 \times 10^{-3} \, M \, CH_3I/\text{min}$

7.13 Reactions involving aqueous solution of ions require no bond breaking and are very fast because of low activation energies. Reactions between covalent molecules require covalent bonds to be broken, requiring higher activation energies, thus slower reaction rates.

7.15 Although not impossible, it is unlikely that one molecule of O_2 and four molecules of ClO_2 will collide together at the same place, with the correct orientation, and sufficient energy to form a transition state and subsequent product.

7.17 The general rule for temperature effect on reaction rates states that for every temperature increase of 10 °C, the reaction rate doubles.
temperature: 10°C $\rightarrow$ 20°C $\rightarrow$ 30°C $\rightarrow$ 40°C $\rightarrow$ 50°C
rate: 16 hr $\rightarrow$ 8 hr $\rightarrow$ 4 hr $\rightarrow$ 2 hr $\rightarrow$ 1 hr
A reaction temperature of 50 °C corresponds to a reaction completion time of 1 hr.

7.19 (1) increase the temperature
(2) increase the concentration of reactants
(3) add a catalyst

7.21 A catalyst increases the rate by providing an alternate reaction pathway of lower activation energy.

7.23 Examples of irreversible reactions include: digesting a piece of candy, the rusting of iron, exploding TNT, and the reaction of sodium or potassium with water.

7.25 (a) $K = \dfrac{[H_2O]^2[O_2]}{[H_2O_2]^2}$

   (b) $K = \dfrac{[N_2O_4]^2[O_2]}{[N_2O_5]^2}$

   (c) $K = \dfrac{[C_6H_{12}O_6][O_2]^6}{[H_2O]^6[CO_2]^6}$

7.27 $K = \dfrac{[CO_2][H_2]}{[H_2O][CO]} = \dfrac{[0.133\ M][3.37\ M]}{[0.72\ M][0.933\ M]} = 0.67$

7.29 $K = \dfrac{[NO]^2[Cl_2]}{[NOCl_2]^2} = \dfrac{[1.4\ M]^2[0.34\ M]}{[2.6\ M]^2} = 0.099\ M$

7.31 K > 1, equilibrium favors products; K < 1, equilibrium favors reactants. products favored: (a), (b), and (c)   reactants favored: (d) and (e)

7.33 (a) No, the rate of reaction is independent of the energy difference between products and reactant. The rate of reaction is inversely proportional to the activation energy.
   (b) Yes, endothermic reactions always have high energy of activations, thus proceed slowly.

7.35 (a) right   (b) right   (c) left   (d) left   (e) no shift

7.37 (a) Adding Br2 (a reactant), will shift the equilibrium to the right.
   (b) The equilibrium constant will remain the same.

7.39 (a) no change   (b) no change   (c) smaller
   Equilibrium constants are independent of reactant and product concentrations. The K of a endothermic reaction will decrease with decreasing temperature

7.41 As temperatures increase, the rates of most chemical processes increase. A high body temperature is dangerous because metabolic processes (including digestion, respiration, and the biosynthesis of essential compounds) take place at a rate faster than what is safe for the body. As temperatures decrease, so do the rates of most chemical reactions. As body temperatures decrease below normal, the vital chemical reactions will slow down to rates slower than what is safe for the body.

7.43 The capsule with the tiny beads will act faster than the solid pill form. The small bead size increases the drug's surface area allowing the drug to react faster and deliver its therapeutic effects more quickly.

7.45 The addition of heat is used to increase the rate of reaction. The addition of a catalyst permits the reaction to take place at a convenient rate and temperature.

7.47 The following energy diagram can be drawn for an exothermic reaction:

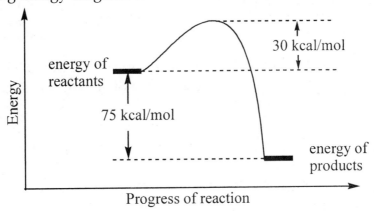

7.49 rate = k[NOBr]

$$k = \frac{rate}{[NOBr]} = \frac{-2.3 \text{ mol NOBr}/L \cdot hr}{6.2 \text{ mol NOBr}/L} = -0.37/hr$$

7.51 $rate = \dfrac{[0.180 \text{ mol/L } N_2O_4 - 0.200 \text{ mol/L } N_2O_4]}{10 \text{ s}} = -2.0 \times 10^{-3} \text{ mol } N_2O_4/L \cdot s$

7.53 Before equilibrium, [α-glucose] = 1 M and [β-glucose] = 0 M.
At equilibrium, [α-glucose] = 1 M - x and [β-glucose] = x

$$1.5 = \frac{[\beta\text{-glucose}]}{[\alpha\text{-glucose}]} = \frac{[x]}{[1\,M - x]}$$

2.5x = 1.5 M

x = 0.6 M

[α-glucose] = 0.4 M and [β-glucose] = 0.6 M

7.55 $4NH_3(g) + 7O_2(g) \rightleftharpoons 4NO_2(g) + 6H_2O(g)$

7.57 Reaction A involving spherical molecules goes faster because for spherical molecules, orientation does not matter for molecular collisions to yield product. Rod-like molecules require special orientations for effective collisions to occur.

7.59 rate = $\dfrac{[0.30 \text{ mol/L } I_2 - 0 \text{ mol/L } I_2]}{10 \text{ s}}$ = 3.0×10^{-2} mol I_2 / L·s

7.61 (a) The rate-determining steps is the lowest step: step 2 (k = 0.05 M).
(b) The fastest step will have the lowest energy of activation: step 3 (k = 4.2 M).

Chapter 8 Acids and Bases

8.1 The following are acid-base equilibria:

(a) H_3O^+ + I^- ⇌ H_2O +   HI
    weaker acid   weaker base   stronger base   stronger acid

(b) CH_3COO^- + H_2S ⇌ CH_3COOH + HS^-
    weaker base   weaker acid   stronger acid   stronger base

8.3 The lower pKa is the stronger acid:
(a) Ascorbic acid (vitamin C), pK_a 4.1
(b) Aspirin, pK_a 3.49

8.5 $pH = -\log[H_3O^+]$ $[H_3O^+] = 10^{-pH}$
(a) $[H_3O^+] = 3.5 \times 10^{-3}$ M; $pH = -\log[3.5 \times 10^{-3}$ M$] = 2.5$
(b) pH of tomato juice is 4.10; $[H_3O^+] = 10^{-4.1} = 7.9 \times 10^{-5}$ M

8.7 The 25.0 mL acetic acid sample was titrated with an average volume of 19.83 mL 0.121 M NaOH solution.

$$19.83 \text{ mL NaOH} \left(\frac{1 \text{ L NaOH}}{1000 \text{ mL NaOH}} \right)\left(\frac{0.121 \text{ mol NaOH}}{1 \text{ L NaOH}} \right) = 2.40 \times 10^{-3} \text{ mol NaOH}$$

$$\text{mol of acetic acid} = 2.40 \times 10^{-3} \text{ mol NaOH} \left(\frac{1 \text{ mol HOAc}}{1 \text{ mol NaOH}} \right) = 2.40 \times 10^{-3} \text{ mol}$$

$$[\text{acetic acid}] = \frac{2.40 \times 10^{-3} \text{ mol acetic acid}}{25.0 \text{ mL solution}} \left(\frac{1000 \text{ mL sol}}{1 \text{ L sol}} \right) = 0.0960 \text{ } M \text{ acetic acid}$$

8.9 Use the Henderson-Hasselbalch Equation: $pH = pK_a + \log[A^-]/[HA]$ where: $pK_a = -\log K_a$ of the weak acid, HA and A^- are the weak acid and conjugate base respectively.

$$pH = -\log(7.3 \times 10^{-10}) + \log\left(\frac{0.50 \text{ } M}{0.25 \text{ } M} \right) = 9.44$$

8.11 Listed below are the following acid ionization equilibrium equations:

(a) $HNO_3(aq) + H_2O(l) \rightleftharpoons H_3O^+(aq) + NO_3^-(aq)$

(b) $HBr(g) + H_2O(l) \rightleftharpoons H_3O^+(aq) + Br^-(aq)$

(c) $H_2SO_3(aq) + H_2O(l) \rightleftharpoons H_3O^+(aq) + HSO_3^-(aq)$

(d) $H_2SO_4(aq) + H_2O(l) \rightleftharpoons H_3O^+(aq) + HSO_4^-(aq)$

(e) $HCO_3^-(aq) + H_2O(l) \rightleftharpoons H_3O^+(aq) + CO_3^{2-}(aq)$

(f) $H_3BO_3(aq) + H_2O(l) \rightleftharpoons H_3O^+(aq) + H_2BO_3^-(aq)$

8.13 Equations involving base ionization in water:

(a) $LiOH(s) + H_2O(l) \longrightarrow Li^+(aq) + OH^-(aq)$

(b) $(CH_3)_2NH(l) + H_2O(l) \longrightarrow (CH_3)_2NH_2^+(aq) + OH^-(aq)$

8.15 A conjugate base is the species that results after the acid lost its proton.

(a) HSO_4^-	(b) $H_2BO_3^-$	(c) I^-
(d) H_2O	(e) NH_3	(f) PO_4^{3-}

8.17 A conjugate acid results from a base acquiring a proton from an acid:

(a) H_2O	(b) H_2S	(c) NH_4^+
(d) CH_3CH_2OH	(e) HCO_3^-	(f) H_2CO_3

8.19 The equilibrium favors the side with the weaker acid-weaker base. Equilibria (a) and (c) lie to the right, equilibrium (b) lies to the left.

(a) H_3PO_4 + OH^- $\rightleftharpoons$ $H_2PO_4^-$ + H_2O

  stronger acid   stronger base   weaker base   weaker acid

(b) H_2O + Cl^- $\rightleftharpoons$ HCl + OH^-

  weaker acid   weaker base   stronger acid   stronger base

(c) HCO_3^- + OH^- $\rightleftharpoons$ CO_3^{2-} + H_2O

  stronger acid   stronger base   weaker base   weaker acid

8.21 Carbonic acid exists in solution as an equilibrium between it and CO_2 and H_2O.

$$H_2CO_3(aq) \rightleftharpoons CO_2(g) + H_2O(l)$$

8.23 (a) Strong acids have smaller pK_a's, therefore weak acids have large pK_a's.
(b) Strong acids have large K_a's.

8.25 At equal concentrations, pH decreases (becomes more acidic) as K_a increases.
(a) 0.10 M HCl   (b) 0.10 M H_3PO_4 (c) 0.010 M H_2CO_3
(d) 0.10 M NaH_2PO_4 (e) 0.10 M aspirin

8.27 Only (b) Mg involves a redox reaction. The other reactions are acid-base reactions.
(a) Na_2CO_3 + 2HCl → CO_2 + 2NaCl + H_2O

(b) Mg + 2HCl → $MgCl_2$ + H_2

(c) NaOH + HCl → NaCl + H_2O

(d) Fe_2O_3 6HCl → $2FeCl_3$ + $3H_2O$

(e) NH_3 + HCl → NH_4Cl

(f) CH_3NH_2 + HCl → CH_3NH_3Cl

(g) $NaHCO_3$ + HCl → H_2CO_3 + NaCl → CO_2 + H_2O + NaCl

8.29 Using the equation: $[H_3O^+][OH^-] = 1.0 \times 10^{-14}\ M^2$
(a) $[OH^-] = 10^{-3}\ M$ (b) $[OH^-] = 10^{-10}\ M$
(c) $[OH^-] = 10^{-7}\ M$ (d) $[OH^-] = 10^{-15}\ M$

8.31 Using the equation: pH = -log$[H_3O^+]$:
(a) pH = 8 (basic)   (b) pH = 10 (basic)   (c) pH = 2 (acidic)
(d) pH = 0 (acidic)   (e) pH = 7 (neutral)

8.33 Using the equation: pH = -log$[H_3O^+]$ and pH + pOH = 14.
(a) pH = 8.5 (basic)   (b) pH = 1.2 (acidic)
(c) pH = 11 (basic)   (d) pH = 6.3 (acidic)

8.35 Using the equation: $[OH^-] = 10^{-pOH}$ and pH + pOH = 14.
(a) pOH = 1.0, $[OH^-]$ = 0.10 M (b) pOH = 2.4, $[OH^-]$ = 4.0 x $10^{-3}\ M$
(c) pOH = 2.0, $[OH^-]$ = 1.0 x $10^{-2}\ M$ (d) pOH = 5.6, $[OH^-]$ = 2.5 x $10^{-6}\ M$

8.37 The following reactions involve the formation of salts:

(a) $Ba(OH)_2 + 2HCl \rightarrow BaCl_2 + 2H_2O$

(b) $NaOH + HNO_3 \rightarrow NaNO_3 + H_2O$

(c) $HCOOH + NH_3 \rightarrow HCOONH_4$

(d) $Ca(OH)_2 + H_2SO_4 \rightarrow CaSO_4 + 2H_2O$

(e) $Mg(OH)_2 + H_2SO_4 \rightarrow MgSO_4 + 2H_2O$

(f) $NaH_2PO_4 + NaOH \rightarrow Na_2HPO_4 + H_2O$

$2NaOH + H_3PO4 \rightarrow Na_2HPO_4 + 2H_2O$

(g) $NH_3 + HCl \rightarrow NH_4Cl$

8.39 When sodium carbonate is dissolved in water, some of the carbonate ion is converted to hydrogen carbonate ions and hydroxide ions, raising the pH to basic values.

8.41 $M = \dfrac{mol}{L} = \left(\dfrac{12.7 \text{ g HCl}}{1.00 \text{ L sol}} \right)\left(\dfrac{1 \text{ mol HCl}}{36.5 \text{ g HCl}} \right) = 0.348 \; M \; HCl$

8.43 (a) 12 g of NaOH diluted to 400 mL of solution:

$$400 \text{ mL sol} \left(\dfrac{1 \text{ L sol}}{1000 \text{ mL sol}} \right)\left(\dfrac{0.75 \text{ mol NaOH}}{1 \text{ L sol}} \right)\left(\dfrac{40.0 \text{ g NaOH}}{1 \text{ mol NaOH}} \right) = 12 \text{ g NaOH}$$

(b) 12 g of $Ba(OH)_2$ diluted to 1.0 L of solution:

$$\dfrac{0.071 \text{ mol Ba(OH)}_2}{1 \text{ L sol}} \left(\dfrac{171.4 \text{ g Ba(OH)}_2}{1 \text{ mol Ba(OH)}_2} \right) = 12 \text{ g Ba(OH)}_2$$

8.45 5.66 mL of 0.740 M H_2SO_4 are required to titrate 27.0 mL of 0.310 M NaOH.

$$27.0 \text{ mL NaOH sol} \left(\dfrac{1 \text{ L sol}}{1000 \text{ mL sol}} \right)\left(\dfrac{0.310 \text{ mol NaOH}}{1 \text{ L sol}} \right) = 8.37 \times 10^{-3} \text{ mol NaOH}$$

$$8.37 \times 10^{-3} \text{ mol NaOH} \left(\dfrac{1 \text{ mol H}_2\text{SO}_4}{2 \text{ mol NaOH}} \right)\left(\dfrac{1 \text{ L H}_2\text{SO}_4 \text{ sol}}{0.740 \text{ mol H}_2\text{SO}_4} \right)\left(\dfrac{1000 \text{ mL sol}}{1 \text{ L sol}} \right) = 5.66 \text{ mL}$$

8.47 Assuming that the base generates one mole of hydroxide per mole of base:

$$22.0 \text{ mL HCl sol} \left(\dfrac{0.150 \text{ mol HCl}}{1000 \text{ mL sol}} \right)\left(\dfrac{1 \text{ mol H}^+}{1 \text{ mol HCl}} \right) = 3.30 \times 10^{-3} \text{ mol H}^+$$

At the end point, 3.30×10^{-3} mol of H^+ added to 3.30×10^{-3} mol of the unknown base.

8.49 In the CH_3COOH/CH_3COO^- buffer solution, the CH_3COO^- is completely ionized while the CH_3COOH is only partially ionized.

(a) $H_3O^+ + CH_3COO^- \rightleftharpoons CH_3COOH + H_2O$ (removal of H_3O^+)

(b) $HO^- + CH_3COOH \rightleftharpoons CH_3COO^- + H_2O$ (removal of OH^-)

8.51 Using the Henderson-Hasselbalch Equation:

$$pH = pKa + \log\frac{[HCOO^-]}{[HCOOH]} = 3.75 + \log\frac{[0.10]}{[0.10]} = 3.75$$

8.53 The buffer capacity is the ability of a buffer to prevent significant changes in pH upon addition of acid or base.

8.55 (a) According to the Henderson-Hasselbalch equation, no change in pH will be observed as long as the ratio of weak acid/conjugate base ratio remains the same.

(b) The buffer capacity increases with increasing amount of weak acid/conjugate base concentrations, therefore, 1.0 mol amounts of each diluted to 1 L would have a greater buffer capacity than 0.1 mol of each diluted to 1 L.

8.57 Investigating the Henderson-Hasselbalch Equation further:

$$pH = pK_a + \log\frac{[A^-]}{[HA]} \qquad \text{when } [A^-] = [HA], \text{ the } \log\frac{[A^-]}{[HA]} = 0$$

the equation reduces to: $pH = pK_a$

8.59 Using the Henderson-Hasselbalch Equation:

$$7.40 = 7.21 + \log\frac{[HPO_4^{2-}]}{[H_2PO_4^-]}$$

$$\frac{[HPO_4^{2-}]}{[H_2PO_4^-]} = 1.55$$

8.61 When 0.10 mol of sodium acetate is added to 0.10 M HCl, the sodium acetate is completely neutralizes the HCl to acetic acid and sodium chloride. The pH of the solution is determined by the incomplete ionization of acetic acid.

$$K_a = \frac{[CH_3COO^-][H_3O^+]}{[CH_3COOH]} \qquad [H_3O^+] = [CH_3COO^-] = x$$

$$\sqrt{x^2} = \sqrt{K_a[CH_3COOH]} = \sqrt{(1.8 \times 10^{-5})(0.10)}$$

$$x = [H_3O^+] = 1.34 \times 10^{-3} \; M$$

$$pH = -\log[H_3O^+] = 2.9$$

8.63 The most important immediate first aid for chemical burns to the eyes is washing them with lots of water, with a subsequent trip to a physician.

8.65 The most common bases used in over-the-counter antacids include $CaCO_3$, $MgCO_3$, and $NaHCO_3$.

8.67 The equilibrium favors the side of the weaker acid/weaker base.
   (a) 4-methyl phenol is soluble in aqueous NaOH.
   $$CH_3C_6H_4OH \; + \; NaOH \; \rightleftharpoons \; CH_3C_6H_4O^- \; + \; H_2O$$
   $pK_a = 10.26$ $\qquad\qquad\qquad\qquad\qquad$ $pK_a = 15.56$
   (b) 4-methyl phenol is insoluble in aqueous $NaHCO_3$.
   $$CH_3C_6H_4OH \; + \; NaHCO_3 \; \rightleftharpoons \; CH_3C_6H_4O^- \; + \; H_2CO_3$$
   $pK_a = 10.26$ $\qquad\qquad\qquad\qquad\qquad$ $pK_a = 6.37$
   (c) 4-methyl phenol is insoluble in aqueous NH_3.
   $$CH_3C_6H_4OH \; + \; NH_3 \; \rightleftharpoons \; CH_3C_6H_4O^- \; + \; NH_4^+$$
   $pK_a = 10.26$ $\qquad\qquad\qquad\qquad\qquad$ $pK_a = 9.25$

8.69 Dilute HCl (0.10 M) is more acidic than concentrated acetic acid (5.0 M) because HCl, being a strong acid, is completely ionized in water to give an $[H_3O^+] = 0.10\ M$ and a pH = 1. Acetic acid is considered a weak acid ($K_a = 1.8 \times 10^{-5}$) and not completely ionized. Using the equation below, 5 M acetic acid ionizes to give an approximate $[H_3O^+] = 0.010\ M$ and a pH of 2.0.

$$K_a = \frac{[CH_3COO^-][H_3O^+]}{[CH_3COOH]} \qquad [H_3O^+] = [CH_3COO^-] = x$$

$$\sqrt{x^2} = \sqrt{Ka[CH_3COOH]} = \sqrt{(1.8 \times 10^{-5})(5.0M)}$$

$$x = [H_3O^+] = 0.010\ M$$

$$pH = -\log[H_3O^+] = 2.0$$

8.71 The solution of oxalic acid is $3.70 \times 10^{-3}\ M$

$$\frac{0.583\ g\ H_2C_2O_4}{1.75\ L\ sol} \left(\frac{1\ mol\ H_2C_2O_4}{90.04\ g\ H_2C_2O_4} \right) = 3.70 \times 10^{-3}\ M\ oxalic\ acid$$

8.73 The concentration of barbituric acid equilibrates to 0.90 M.

$$K_a = \frac{[barbiturate^-][H_3O^+]}{[barbituric\ acid]} \qquad x = [H_3O^+] = [barbiturate^-]$$

$$[barbituric\ acid] = \frac{x^2}{K_a} = \frac{(0.0030)^2}{1.0 \times 10^{-5}} = 0.90\ M$$

8.75 Yes, a pH = 0 is possible. A 1.0 M solution of HCl has a $[H_3O^+] = 1.0\ M$.
   $pH = -\log[H_3O^+] = -\log[1.0\ M] = 0$

8.77 A 1.0 M solution of CH_3COOH and 1.0 M solution of HCl do not have the same pH. An aqueous solution of HCl is a strong acid that completely ionizes, thus the 1.0 M HCl yields 1.0 M H_3O^+ ions. An aqueous 1.0 M solution CH_3COOH is a weak acid that does not completely ionize, thus giving a fraction of 1.0 M H_3O^+ ions in solution.

8.79 Using the Henderson-Hasselbalch equation:

$$\frac{[H_2BO_3^-]}{[H_3BO_3]} = 10^{pH-pKa} = 10^{8.40-9.14}$$

$$\frac{[H_2BO_3^-]}{[H_3BO_3]} = 0.182$$

Need 0.182 mol of $H_2BO_3^-$ and 1.00 mol of H_3BO_3 in 1.00 L of solution.

8.81 An equilibrium will favor the side of the weaker acid/weaker base. The larger the pKa value, the weaker the acid.

8.83 (a) $HCOO^-$ + H_3O^+ $\rightleftharpoons$ $HCOOH$ + H_2O

(b) $HCOOH$ + HO^- $\rightleftharpoons$ $HCOO^-$ + H_2O

8.85 Using the Henderson-Hasselbalch Equation:

(a) $[Na_2HPO_4] = [NaH_2PO_4]10^{pH-pKa} = [0.050M]10^{7.21-7.21} = 0.050\ M$

(b) $[Na_2HPO_4] = [NaH_2PO_4]10^{pH-pKa} = [0.050M]10^{6.21-7.21} = 0.0050\ M$

(c) $[Na_2HPO_4] = [NaH_2PO_4]10^{pH-pKa} = [0.050M]10^{8.21-7.21} = 0.50\ M$

8.87 According to the Henderson-Hasselbalch equation:

$$pH = 7.21 + \log\frac{[HPO_4^{2-}]}{[H_2PO_4^-]}$$

As the concentration of $H_2PO_4^-$ increases, the $\log\frac{[HPO_4^{2-}]}{[H_2PO_4^-]}$ becomes negative,

thus lowering the pH and becoming more acidic.

Chapter 9 Nuclear Chemistry

9.1 $^{139}_{53}I \rightarrow ^{0}_{-1}e + ^{139}_{54}Xe$

9.3 $^{74}_{33}As \rightarrow ^{0}_{+1}e + ^{74}_{32}Ge$

9.5 The intensity of any radiation decreases with the square of the distance: $\dfrac{I_1}{I_2} = \dfrac{d_2^{\,2}}{d_1^{\,2}}$

$$\frac{300 \text{ mCi}}{I_2} = \frac{(3.0 \text{ m})^2}{(0.01 \text{ m})^2}$$

$$I_2 = \frac{(300 \text{ mCi})(0.01 \text{ m})^2}{(3.0 \text{ m})^2} = 3.3 \times 10^{-3} \text{ mCi}$$

9.7 Yes, all types of electromagnetic radiation travel at the speed of light in a vacuum. Gamma rays are an energetic form of electromagnetic radiation.

9.9 $f = \dfrac{c}{\lambda} = \dfrac{3.0 \times 10^{10} \text{ cm/s}}{5.8 \text{ cm}} = 5.2 \times 10^{9}/\text{s}$

9.11 $f = \dfrac{c}{\lambda} = \dfrac{3.0 \times 10^{8} \text{ m/s}}{650 \text{ nm}} \left(\dfrac{10^{9} \text{ nm}}{1 \text{ m}} \right) = 4.6 \times 10^{14}/\text{s}$

9.13 (a) $^{19}_{9}F$ (b) $^{32}_{15}P$ (c) $^{87}_{37}Rb$

9.15 Boron-10 is the most stable isotope because of an equal number of protons and neutrons in the nucleus.

9.17 (a) $^{159}_{63}Eu \rightarrow ^{0}_{-1}e + ^{159}_{64}Gd$
(b) $^{141}_{56}Ba \rightarrow ^{0}_{-1}e + ^{141}_{57}La$
(c) $^{242}_{95}Am \rightarrow ^{0}_{-1}e + ^{242}_{96}Cm$

9.19 (a) $^{210}_{83}Bi \rightarrow ^{4}_{2}He + ^{206}_{81}Tl$
(b) $^{238}_{94}Pu \rightarrow ^{4}_{2}He + ^{234}_{92}U$
(c) $^{174}_{72}Hf \rightarrow ^{4}_{2}He + ^{170}_{70}Yb$

9.21 $^{29}_{15}P \rightarrow {}^{0}_{+1}e + {}^{29}_{14}Si$

9.23 $^{238}_{92}U \rightarrow {}^{4}_{2}He + {}^{234}_{90}Th$ alpha emission

$^{234}_{90}Th \rightarrow {}^{0}_{-1}e + {}^{234}_{91}Pa$ beta emission

$^{234}_{91}Pa \rightarrow {}^{0}_{-1}e + {}^{234}_{92}U$ beta emission

9.25 (a) $^{16}_{8}O + {}^{16}_{8}O \rightarrow {}^{4}_{2}He + {}^{28}_{14}Si$

(b) $^{235}_{92}U + {}^{1}_{0}n \rightarrow {}^{90}_{38}Sr + {}^{143}_{54}Xe + 3\,{}^{1}_{0}n$

(c) $^{13}_{6}C + {}^{4}_{2}He \rightarrow {}^{16}_{8}O + {}^{1}_{0}n$

(d) $^{210}_{83}Bi \rightarrow {}^{0}_{-1}e + {}^{210}_{84}Po$

(e) $^{12}_{6}C + {}^{1}_{1}H \rightarrow {}^{13}_{7}N + \gamma$

9.27 After three half-lives: $1/2 \times 1/2 \times 1/2 = 1/8$ or 12.5% of the original amount remains.

9.29 No, the conversion of Ra to Ra^{2+} involves the loss of valence electrons, which is not a nuclear process and does not involve a change in radioactivity.

9.31 50.0 mg $\rightarrow$ 25.0 mg $\rightarrow$ 12.5 mg $\rightarrow$ 6.25 mg $\rightarrow$ 3.12 mg
4 half-lives in 60 minutes: half-life = 60 min/4 = 15 minutes

9.33 No. Scintillation counters count each particle one by one and emit light when each particle strikes the phosphor.

9.35 Geiger counters measure the (a) intensity of radiation.

9.37 (a) amount of radiation absorbed by tissues
(b) effective dose absorbed by humans
(c) effective does delivered
(d) intensity of radiation
(e) amount of radiation absorbed by tissues
(f) intensity of radiation
(g) effective dose absorbed by humans

9.39 Alpha particles have so little penetrating power that they are stopped by the thick layer of skin on the hand. If they get into the lung, the thin membranes offer little resistance to the particles, which then damage the cells of the lung.

9.41 Alpha particles are the most damaging to tissue.

9.43 (a) iodine-125 (prostate cancer), iodine-131 (thyroid cancer), and cobalt-60 (general cancer treatment)
(b) selenium-75
(c) strontium-85
(d) carbon-11 and technetium-99m
(e) mercury-197 and technetium-99m

9.45 The half-life is too short. It would not last long enough to be delivered to the hospital.

9.47 The most abundant nucleus in the universe is $_1^1 H$.

9.49 $_{96}^{248} Cm + _2^4 He \rightarrow 2 _0^1 n + _1^1 H + _{97}^{249} Bk$

9.51 $_{82}^{208} Pb + _{36}^{86} Kr \rightarrow 4 _0^1 n + _{118}^{290} Unknown$

9.53 Gamma radiation is not deflected by charged plates because they are uncharged.

9.55 2003 – 1350 = 653 years (if the experiment was run in the year 2003)
653 years/5730 = 0.114 half-lives

9.57 Radon-222 produced polonium-218 by alpha emission:
$_{86}^{222} Rn \rightarrow _2^4 He + _{84}^{218} Po$

9.59 An MRI does not use ionizing radiation as an imaging tool. Bones are also transparent to MRI, enabling a clearer analysis of soft tissue.

9.61 Positrons quickly collide with readily available electrons in matter, and then annihilate each other, resulting in gamma radiation.

9.63 Iodine-131 is concentrated in the thyroid where the radiation can induce thyroid cancer.

9.65 $_{10}^{19} Ne \rightarrow _{+1}^0 e + _9^{19} F$
$_{11}^{20} Na \rightarrow _{+1}^0 e + _{10}^{20} Ne$

9.67 Both the curie and the becquerel have units of disintegrations/second, a measurement of radiation intensity.

9.69 (a) $\dfrac{294 \text{ mrem/yr}}{359 \text{ mrem/yr}} \times 100 = 82\%$ (b) $\dfrac{39 \text{ mrem/yr}}{359 \text{ mrem/yr}} \times 100 = 11\%$

(c) $\dfrac{0.5 \text{ mrem/yr}}{359 \text{ mrem/yr}} \times 100 = 0.1\%$

9.71 X-rays will cause more ionization than radar because X-Rays are higher energy.

9.73 $1000/475 = 2$ half-lives: $1/2 \times 1/2 = 1/4$ so 25% of the original Americium will be around after 1000 years.

9.75 One sievert is equal to 100 rem. This is sufficient to cause radiation sickness but not certain death.

9.77 (a) Radioactive elements are constantly decaying to other isotopes and elements mixed in with the original isotopes.
(b) Beta emissions result from the decay of a neutron in the nucleus to a proton (the increase in atomic number) and an electron (beta particle).

9.79 Oxygen-16 is stable because it has an equal number of protons and neutrons. The others are unstable because the numbers of protons and neutrons are unequal. In this case, the greater the difference in numbers of protons and neutrons, the faster the isotope decays.

9.81 $^{208}_{82}\text{Pb} + ^{64}_{28}\text{Ni} \rightarrow 6\,^{1}_{0}\text{n} + ^{266}_{110}\text{X}$

Chapter 10 Organic Chemistry

10.1 Following are Lewis structures showing all bond angles.

(a) $H-\underset{\substack{H\\|}}{\overset{\substack{H\\|}}{C}}-\underset{\substack{H\\|}}{\overset{\substack{H\\|}}{C}}-\overset{\cdot\cdot}{\underset{\cdot\cdot}{O}}-H$ 109.5°, 109.5°

(b) $H-\underset{\substack{H\\|}}{\overset{\substack{H\\|}}{C}}-\underset{\substack{H\\|}}{\overset{\substack{H\\|}}{C}}=\underset{\substack{|\\H}}{\overset{\substack{H\\|}}{C}}-H$ 109.5°, 120°

10.2 Of the four alcohols with molecular formula $C_4H_{10}O$, two are 1°, one is 2°, and one is 3°. For the Lewis structures of the 3° alcohol and two of the 1° alcohols, some C-CH₃ bonds are drawn longer to avoid crowding in the formulas.

$H-\underset{\substack{H\\|}}{\overset{\substack{H\\|}}{C}}-\underset{\substack{H\\|}}{\overset{\substack{H\\|}}{C}}-\underset{\substack{H\\|}}{\overset{\substack{H\\|}}{C}}-\underset{\substack{H\\|}}{\overset{\substack{H\\|}}{C}}-\overset{\cdot\cdot}{\underset{\cdot\cdot}{O}}-H$ $CH_3CH_2CH_2CH_2OH$
Primary (1°)

$H-\underset{\substack{H\\|}}{\overset{\substack{H\\|}}{C}}-\underset{\substack{H\\|}}{\overset{\substack{H\\|}}{C}}-\underset{\substack{H\\|}}{\overset{\substack{\cdot\cdot\\O\\\cdot\cdot}}{C}}-\underset{\substack{H\\|}}{\overset{\substack{H\\|}}{C}}-H$ $CH_3CH_2\overset{\substack{OH\\|}}{C}HCH_3$
Secondary (2°)

$H-\underset{\substack{H\\|}}{\overset{\substack{|\\C\\|}}{C}}-\underset{\substack{H\\|}}{\overset{\substack{H\\|}}{C}}-\overset{\cdot\cdot}{\underset{\cdot\cdot}{O}}-H$ $CH_3\overset{\substack{CH_3\\|}}{C}HCH_2OH$
Primary (1°)

$H-\underset{\substack{|\\C\\|}}{\overset{\substack{|\\C\\|}}{C}}-\overset{\cdot\cdot}{\underset{\cdot\cdot}{O}}H$ $CH_3\overset{\substack{CH_3\\|}}{\underset{\substack{|\\CH_3}}{C}}OH$
Tertiary (3°)

10.3 The three secondary (2°) amines with the molecular formula $C_4H_{11}N$ are

$CH_3CH_2CH_2NHCH_3$ $CH_3\overset{\substack{CH_3\\|}}{C}HNHCH_3$ $CH_3CH_2NHCH_2CH_3$

10.4 The three ketones with the molecular formula $C_5H_{10}O$ are

$CH_3CH_2CH_2\overset{\substack{O\\||}}{C}CH_3$ $CH_3CH_2\overset{\substack{O\\||}}{C}CH_2CH_3$ $CH_3\overset{\substack{O\\||}}{C}\overset{\substack{\\|}}{\underset{\substack{|\\CH_3}}{C}}HCH_3$

10.5 The two carboxylic acids with the molecular formula $C_4H_8O_2$ are

$CH_3CH_2CH_2\overset{\substack{O\\||}}{C}OH$ $CH_3\overset{\substack{\\|}}{\underset{\substack{|\\CH_3}}{C}}H\overset{\substack{O\\||}}{C}OH$

10.7 For hydrogen, the number of valence electrons plus the number of bonds equals 2. For carbon, nitrogen, and oxygen, the number of valence electrons plus the number of bonds equals 8. Carbon, with four valence electrons, forms four bonds. Nitrogen, with five valence electrons, forms three bonds. Oxygen, with six valence electrons, forms two bonds.

10.9 Following are Lewis structures for each compound.

$$\text{(a)} \quad H\text{-}\underset{\underset{H}{|}}{\overset{\overset{H}{|}}{C}}\text{-}\ddot{O}\text{-}\underset{\underset{H}{|}}{\overset{\overset{H}{|}}{C}}\text{-}H \qquad \text{(b)} \quad H\text{-}\underset{\underset{H}{|}}{\overset{\overset{H}{|}}{C}}\text{-}\underset{\underset{H}{|}}{\overset{\overset{H}{|}}{C}}\text{-}H \qquad \text{(c)} \quad \underset{H}{\overset{H}{}}C\text{=}C\underset{H}{\overset{H}{}}$$

$$\text{(d)} \quad H\text{-}C\equiv C\text{-}H \qquad \text{(e)} \quad \ddot{O}\text{=}C\text{=}\ddot{O} \qquad \text{(f)} \quad \underset{H}{\overset{H}{}}C\text{=}\ddot{O}$$

$$\text{(g)} \quad H\text{-}\ddot{O}\text{-}\overset{\overset{:O:}{\parallel}}{C}\text{-}\ddot{O}\text{-}H \qquad \text{(h)} \quad H\text{-}\underset{\underset{H}{|}}{\overset{\overset{H}{|}}{C}}\text{-}\overset{\overset{:O:}{\parallel}}{C}\text{-}\ddot{O}\text{-}H$$

10.11 In stable organic compounds, carbon must have four covalent bonds to other atoms. In (a), carbon has bonds to five other atoms. In (b), one carbon has four bonds to other atoms, but the second carbon has five bonds to other atoms.

10.13 You would find three regions of electron density and, therefore, predict 120° for the H-N-H bond angles.

10.15 (a) 109.5° about C and O   (b) 120° about C   (c) 180° about C

10.17 (a) Predict 120° for all H-C-C and C-C-C bond angles. (b) planar

10.19 Living organisms use the substances they take in from their environment to produce organic compounds. They also use them to produce inorganic compounds - as for example, the calcium-containing compounds of bones, teeth, and shells.

10.21 Carbon, which forms four bonds; hydrogen, which forms one bond; nitrogen, which forms three bonds and has one unshared pair of electrons; and oxygen, which forms two bonds and has two unshared pairs of electrons.

10.23 A functional group is a part of an organic molecule that undergoes chemical reactions.

Chapter 10 Organic Chemistry

10.25 (a) $-\overset{\overset{\displaystyle :O:}{\|}}{C}-$ (b) $-\overset{\overset{\displaystyle :O:}{\|}}{C}-\overset{..}{\underset{..}{O}}-H$ (c) $-\overset{..}{\underset{..}{O}}-H$ (d) $-\overset{\displaystyle |}{\underset{\displaystyle H}{N}}-H$

10.27 (a) Incorrect. The carbon on the left has five bonds.
(b) Incorrect. The carbon bearing the Cl has five bonds.
(c) Correct.
(d) Incorrect. Oxygen has three bonds and one carbon has five bonds.
(e) Correct.
(f) Incorrect. Carbon on the right has five bonds.

10.29 The one tertiary alcohol with the molecular formula $C_4H_{10}O$ is

$$CH_3\overset{\overset{\displaystyle CH_3}{|}}{\underset{\underset{\displaystyle CH_3}{|}}{C}}OH$$

10.31 The one tertiary amine with the molecular formula $C_4H_{11}N$ is

$$CH_3CH_2\overset{\overset{\displaystyle CH_3}{|}}{N}CH_3$$

10.33 (a) a ketone   (b) two carboxylic acids
(c) two primary amines and a carboxylic acid
(d) two primary alcohols and one ketone

10.35 (a) The four primary alcohols with the molecular formula $C_5H_{12}O$ are

$$CH_3CH_2CH_2CH_2CH_2OH \quad CH_3\overset{\overset{\displaystyle }{|}}{\underset{\underset{\displaystyle CH_3}{|}}{C}}HCH_2CH_2OH \quad CH_3CH_2\overset{\overset{\displaystyle }{|}}{\underset{\underset{\displaystyle CH_3}{|}}{C}}HCH_2OH \quad CH_3\overset{\overset{\displaystyle CH_3}{|}}{\underset{\underset{\displaystyle CH_3}{|}}{C}}CH_2OH$$

(b) The three secondary alcohols with the molecular formula $C_5H_{12}O$ are

$$CH_3\overset{\overset{\displaystyle OH}{|}}{C}HCH_2CH_2CH_3 \qquad CH_3CH_2\overset{\overset{\displaystyle OH}{|}}{C}HCH_2CH_3 \qquad CH_3\overset{\overset{\displaystyle OH}{|}}{C}H\overset{\overset{\displaystyle }{|}}{\underset{\underset{\displaystyle CH_3}{|}}{C}}HCH_3$$

(c) The one tertiary alcohol with the molecular formula $C_5H_{12}O$ is

$$CH_3CH_2\overset{\overset{\displaystyle CH_3}{|}}{\underset{\underset{\displaystyle CH_3}{|}}{C}}OH$$

10.37 The eight carboxylic acids with the molecular formula $C_6H_{12}O_2$ are

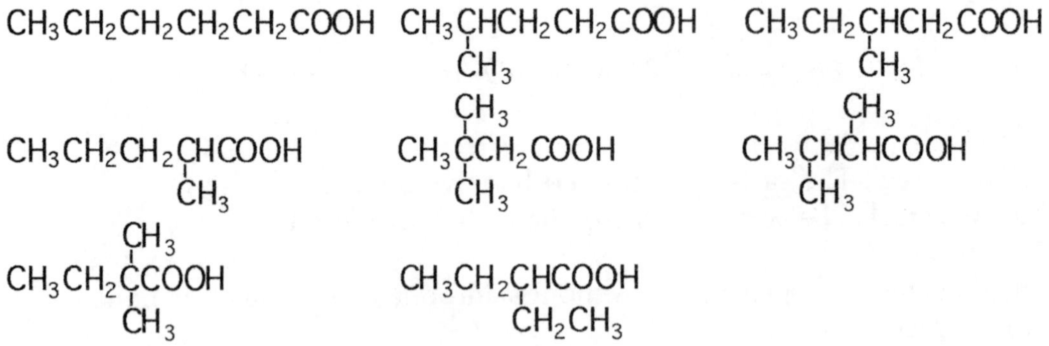

$CH_3CH_2CH_2CH_2CH_2COOH$ $CH_3\underset{\underset{CH_3}{|}}{C}HCH_2CH_2COOH$ $CH_3CH_2\underset{\underset{CH_3}{|}}{C}HCH_2COOH$

$CH_3CH_2CH_2\underset{\underset{CH_3}{|}}{C}HCOOH$ $CH_3\underset{\overset{CH_3}{|}}{\underset{\underset{CH_3}{|}}{C}}CH_2COOH$ $CH_3\underset{\overset{CH_3}{|}}{\underset{\underset{CH_3}{|}}{C}}HCHCOOH$

$CH_3CH_2\underset{\overset{CH_3}{|}}{\underset{\underset{CH_3}{|}}{C}}COOH$ $CH_3CH_2\underset{\underset{CH_2CH_3}{|}}{C}HCOOH$

10.39 Taxol was discovered by a survey of indigenous plants sponsored by the National Cancer Institute with the goal of discovering new chemicals for fighting cancer.

10.41 The goal of combinatorial chemistry is to synthesize large numbers of closely related compounds in one reaction mixture.

10.43 Predict 109.5° for each C-Si-C bond angle.

10.45 (a) CH_3CH_2OH (b) $CH_3CH_2\overset{\overset{O}{\|}}{C}H$ (c) $CH_3\overset{\overset{O}{\|}}{C}CH_3$ (d) $CH_3CH_2\overset{\overset{O}{\|}}{C}OH$

10.47 The three tertiary amines with the molecular formula $C_5H_{13}N$ are

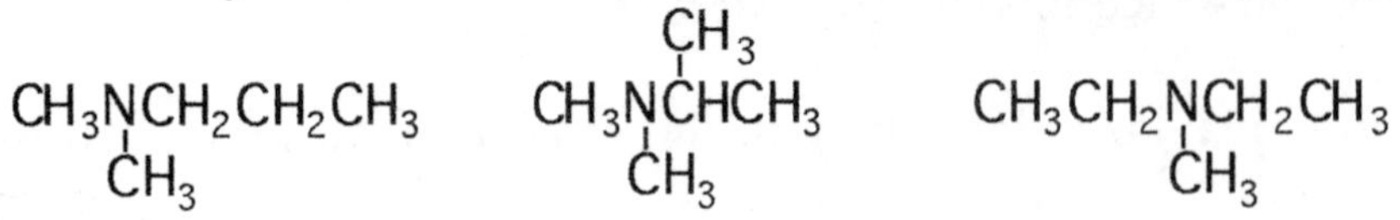

$CH_3\underset{\underset{CH_3}{|}}{N}CH_2CH_2CH_3$ $CH_3\underset{\overset{CH_3}{|}}{\underset{\underset{CH_3}{|}}{N}}CHCH_3$ $CH_3CH_2\underset{\underset{CH_3}{|}}{N}CH_2CH_3$

10.49 (a) O-H is the most polar bond. (b) C-C and C=C are the least polar bonds.

Chapter 11 Alkanes and Cycloalkanes

<u>11.1</u> The alkane is octane, and its molecular formula is C_8H_{18}.

<u>11.2</u> (a) constitutional isomers (b) the same compound

<u>11.3</u> The three constitutional isomers with the molecular formula C_5H_{12} are

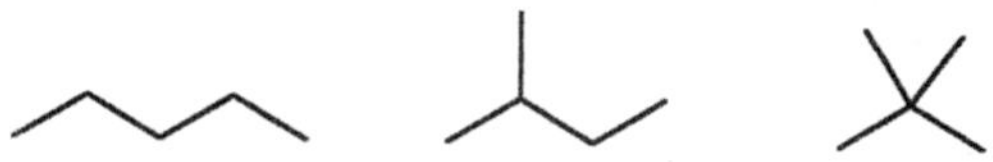

<u>11.4</u> (a) 5-isopropyl-2-methyloctane, $C_{12}H_{26}$
(b) 4-isopropyl-4-propyloctane, $C_{14}H_{30}$

<u>11.5</u> (a) isobutylcyclopentane, C_9H_{18} (b) *sec*-butylcycloheptane, $C_{11}H_{22}$
(c) 1-ethyl-1-methylcyclopropane, C_6H_{12}

<u>11.6</u> The structure with the three methyl groups equatorial is

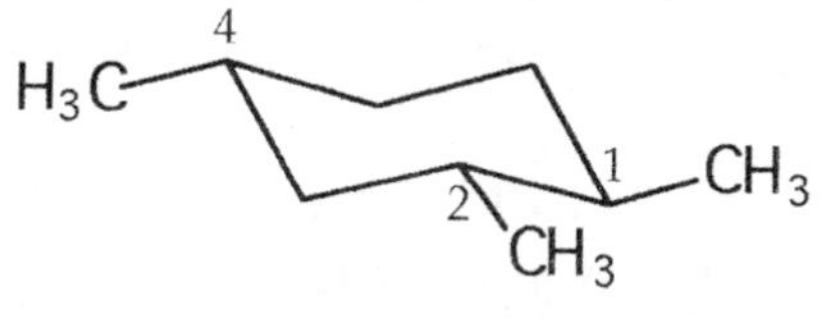

<u>11.7</u> Cycloalkanes (a) and (c) show cis-trans isomerism.

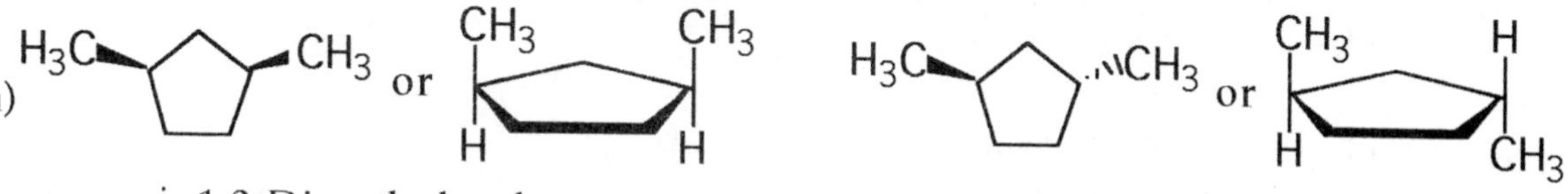

cis-1,3-Dimethylcyclopentane   *trans*-1,3-Dimethylcyclopentane

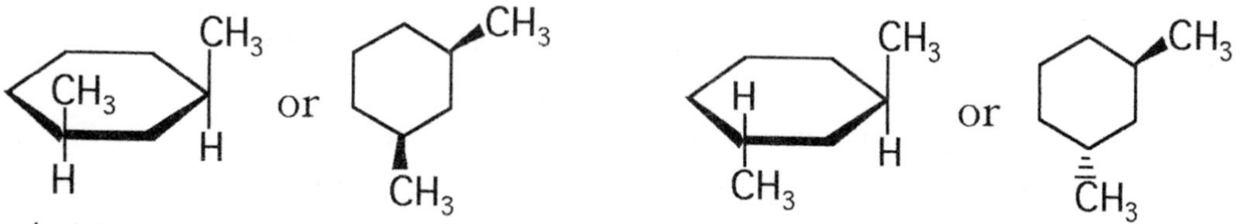

cis-1,3-Dimethylcyclohexane   *trans*-1,3-Dimethylcyclohexane

<u>11.8</u> In order of increasing boiling point, they are:
(a) 2,2-dimethylpropane (9.5°C), 2-methylbutane (27.8°C), pentane (36.1°C)
(b) 2,2,4-trimethylhexane, 3,3-dimethylheptane, nonane

<u>11.9</u> The two products with the formula C_3H_7Cl are

63

Chapter 11 Alkanes and Cycloalkanes

$$CH_3CH_2CH_2Cl$$
1-Chloropropane
(Propyl chloride)

$$CH_3\overset{\overset{\displaystyle Cl}{|}}{C}HCH_3$$
2-Chloropropane
(Isopropyl chloride)

<u>11.11</u> The carbon chain of an alkane is not straight; it is bent with all C-C-C bond angles of approximately 109.5°.

<u>11.13</u> Line-angle formulas are

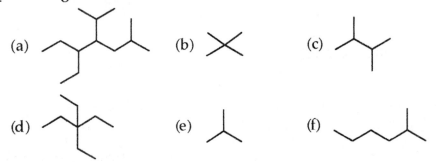

<u>11.15</u> Statements (a) and (b) are true. Statements (c) and (d) are false.

<u>11.17</u> Structures (a) and (g) represent the same compound. Structures (a,g), (c), (d), (e), and (f) represent constitutional isomers.

<u>11.19</u> The structural formulas in parts (b), (c), (e), and (f) represent pairs of constitutional isomers.

<u>11.21</u> (a) ethyl   (b) isopropyl   (c) isobutyl   (d) *tert*-butyl

<u>11.23</u> (a) 2-methylpentane   (b) 2,5-dimethylhexane
(c) 3-ethyloctane   (d) 1-isopropyl-2-methylcyclohexane
(e) isobutylcyclopentane   (f) 1-ethyl-2,4-dimethylcyclohexane

<u>11.25</u> A conformation is any three-dimensional arrangement of the atoms in a molecule that results from rotation about a single bond.

<u>11.27</u> Here are ball-and-stick models of two conformations of ethane. In the staggered conformation, the hydrogen atoms on one carbon are as far apart as possible from the hydrogens on the other carbon. In the eclipsed conformation, they are as close as possible.

64

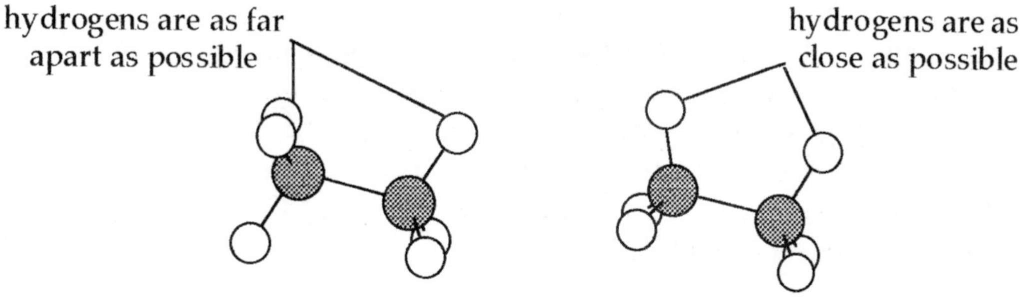

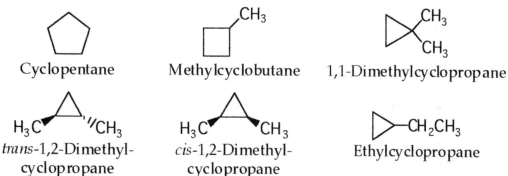

hydrogens are as far apart as possible

hydrogens are as close as possible

staggered conformation   eclipsed conformation

11.29 No.

11.31 Structural formulas for the six cycloalkanes with the molecular formula C_5H_{10} are

Cyclopentane

CH_3

Methylcyclobutane

CH_3
CH_3

1,1-Dimethylcyclopropane

H_3C $''CH_3$

trans-1,2-Dimethyl-
cyclopropane

H_3C CH_3

cis-1,2-Dimethyl-
cyclopropane

CH_2CH_3

Ethylcyclopropane

11.33 The first two representations of menthol show the cyclohexane ring as a planar hexagon. The third shows it as a chair conformation. Notice that in the chair conformation, all three substituents are equatorial.

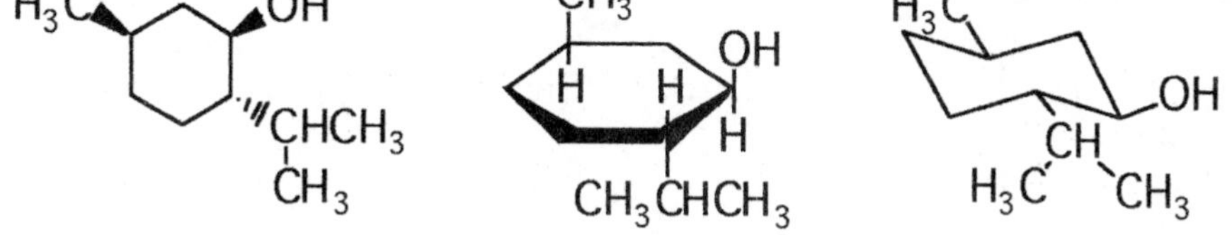

H_3C OH
$'''CHCH_3$
CH_3

CH_3
H   H   OH
                H
CH_3CHCH_3

H_3C
                OH
        CH
H_3C CH_3

11.35 The alternative representation is

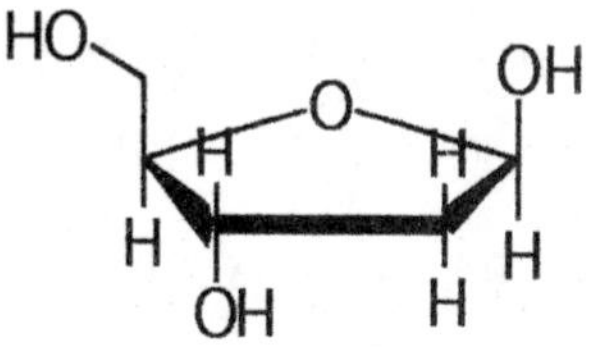

HO
        O   OH
    H   H
H   H
    OH   H

11.37 Heptane, C_7H_{16}, has a boiling point of 98°C and a molecular weight of 100. Its molecular weight is approximately 5.5 times that of water. Although considerably smaller, water molecules associate in the liquid by relatively strong hydrogen bonds whereas the much larger heptane molecules associate only by relatively weak London dispersion forces.

11.39 Alkanes are insoluble in water.

11.41 Boiling points of unbranched alkanes are related to their surface area; the larger the surface area, the greater the strength of dispersion forces, and the higher the boiling point. The relative increase in size per CH_2 group is greatest between CH_4 and CH_3CH_3 and becomes progressively smaller as the molecular weight increases. Therefore, the increase in boiling point per added CH_2 group is greatest between CH_4 and CH_3CH_3, and becomes progressively smaller for higher alkanes.

11.43 Balanced equations for the complete combustion of each are
(a) $2CH_3(CH_2)_4CH_3 + 19O_2 \longrightarrow 12CO_2 + 14H_2O$
    Hexane

(b) 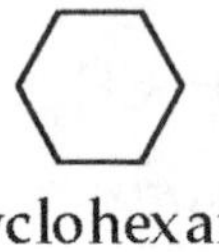 $+ 9O_2 \longrightarrow 6CO_2 + 6H_2O$

    Cyclohexane

(c) $2CH_3\overset{\overset{\displaystyle CH_3}{|}}{C}HCH_2CH_3CH_3 + 19O_2 \longrightarrow 12CO_2 + 14H_2O$
    2-Methylpentane

11.45 Structural formulas for each part are

(a) CH_3Br (b) 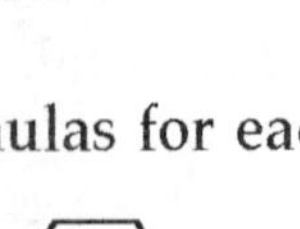 $-Cl$ (c) $BrCH_2CH_2Br$ (d) $CH_3\overset{\overset{\displaystyle CH_3}{|}}{\underset{\underset{\displaystyle Cl}{|}}{C}}CH_3$ (e) CCl_2F_2

11.47 (a) Only one ring contains just carbon atoms. (b) One ring contains two nitrogen atoms. (c) One ring contains two oxygen atoms.

11.49 The presence of Freons in the stratosphere results in destruction of the stratospheric ozone layer.

11.51 An octane rating indicates the relative smoothness with which a gasoline blend burns in an automobile engine. The higher the octane rating, the less the engine knocks. The reference hydrocarbons are 2,2,4-trimethylpentane (isooctane), which is assigned an octane rating of 100, and heptane, which is assigned an octane rating of 0.

<u>11.53</u> 2,2,4-Trimethylpentane has a higher heat of combustion both per mole and per gram.

	Molar mass (g/mol)	Heat of Combustion	
		kcal/mol	kcal/g
2,2,4-Trimethylpentane	114.2	1304	11.4
Ethanol	46.0	327	7.11

<u>11.55</u> (a) The longest chain is pentane. Its IUPAC name is 2-methylpentane.
    (b) The pentane chain is numbered incorrectly. Its IUPAC name is 2-methylpentane.
    (c) The longest chain is pentane. Its IUPAC name is 3-ethyl-3-methylpentane.
    (d) The longest chain is hexane. Its IUPAC name is 3,4-dimethylhexane.
    (e) The longest chain is heptane. Its IUPAC name is 4-methylheptane.
    (f) The longest chain is octane. Its IUPAC name is 3-ethyl-3-methyloctane.
    (g) The ring is numbered incorrectly. Its IUPAC name is 1,1-dimethylcyclopropane.
    (h) The ring is numbered incorrectly. Its IUPAC name is 1-ethyl-3-methylcyclohexane.

<u>11.57</u> Tetradecane is a liquid at room temperature.

<u>11.59</u> Water, a polar molecule, cannot penetrate the surface layer created by this nonpolar hydrocarbon.

Chapter 12 Alkenes and Alkynes

<u>12.1</u> (a) 3,3-dimethyl-1-pentene   (b) 2,3-dimethyl-2-butene
    (c) 3,3-dimethyl-1-butyne

<u>12.2</u> (a) *trans*-3,4-dimethyl-2-pentene   (b) *cis*-4-ethyl-3-heptene

<u>12.3</u> (a) 1-isopropyl-4-methylcyclohexene   (b) cyclooctene
    (c) 4-*tert*-butylcyclohexene

<u>12.4</u> Line-angle formulas for the other two dienes are

cis,trans-2,4-Heptadiene 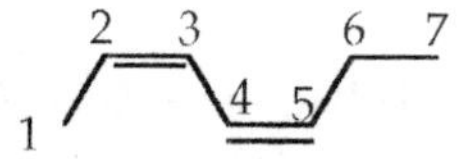   *cis,cis*-2,4-Heptadiene

<u>12.5</u> Four cis-trans isomers are possible.

<u>12.6</u> (a) $CH_3\overset{Br}{\underset{|}{C}}HCH_3$ (b) cyclohexane with Br and CH₃

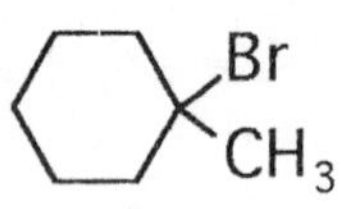

<u>12.7</u> Propose a two-step mechanism similar to that for the addition of HCl to propene. Step 1: Reaction of H^+ with the carbon-carbon double bond gives a 3° carbocation intermediate.

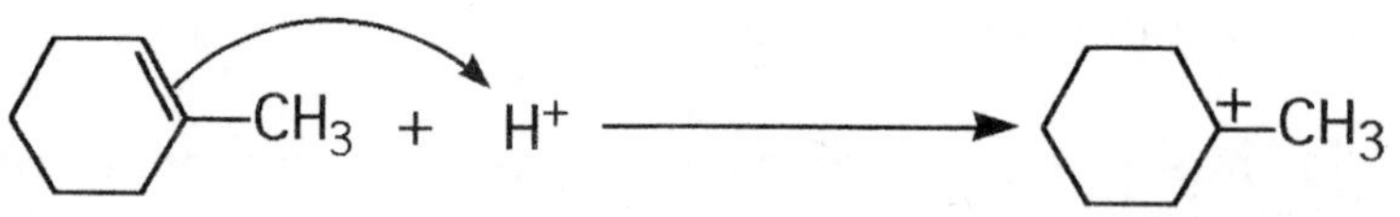

cyclohexene—CH₃ + H⁺ ⟶ cyclohexyl cation ⁺—CH₃

A 3° carbocation
intermediate

Step 2: Reaction of the 3° carbocation intermediate with bromide ion completes the valence shell of carbon and gives the product.

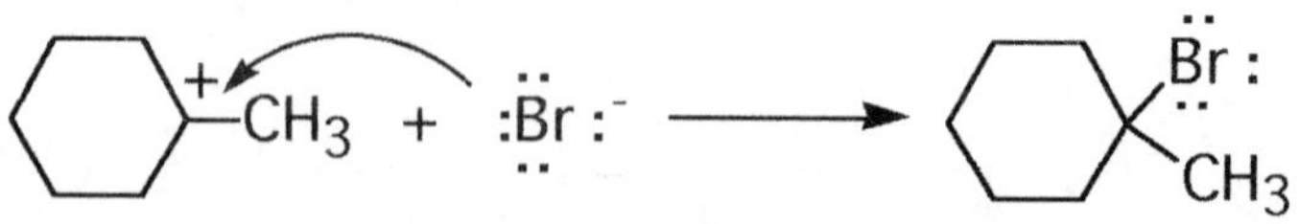

⁺—CH₃ + :Br:⁻ ⟶ Br:—CH₃

<u>12.8</u> The product from each acid-catalyzed hydration is the same alcohol.

$$CH_3\overset{CH_3}{\underset{OH}{C}}CH_2CH_3$$

12.9 Propose a three-step mechanism similar to that for the acid-catalyzed hydration of propene.
Step 1: Reaction of the carbon-carbon double bond with H⁺ gives a 3° carbocation intermediate.

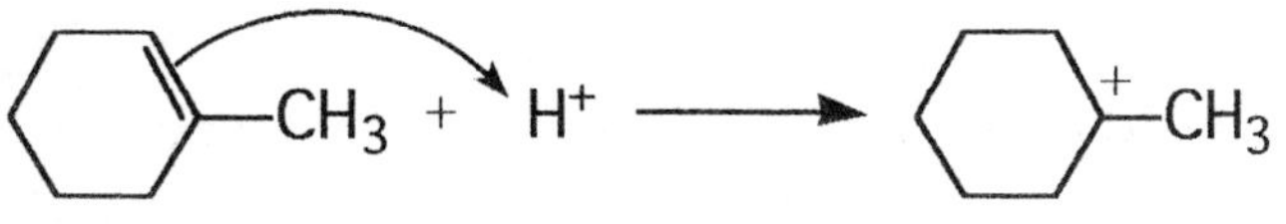

A 3° carbocation intermediate

Step 2: Reaction of the 3° carbocation intermediate with water completes the valence shell of carbon and gives an oxonium ion.

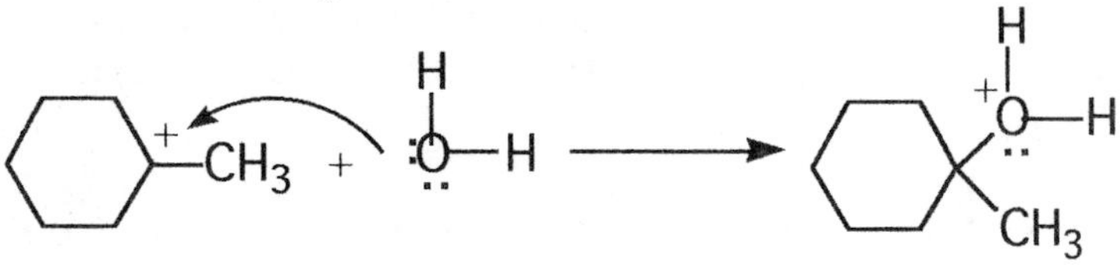

An oxonium ion

Step 3: Loss of H⁺ from the oxonium ion completes the reaction and generates a new H⁺ catalyst.

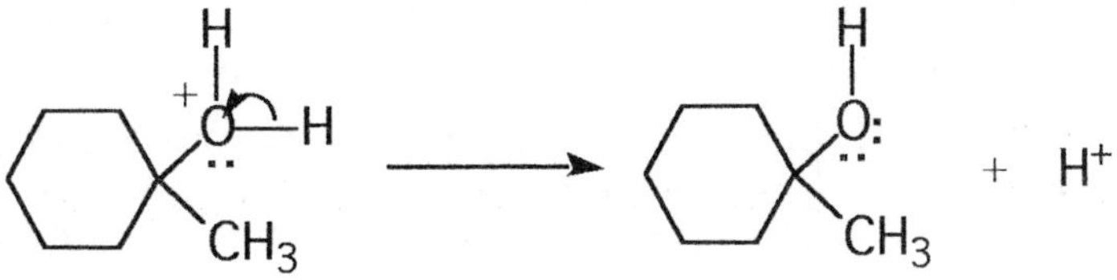

12.10 (a) 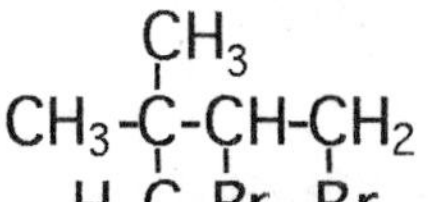 $CH_3\text{-}\underset{H_3C}{\overset{CH_3}{C}}\text{-}\underset{Br}{CH}\text{-}\underset{Br}{CH_2}$ (b)

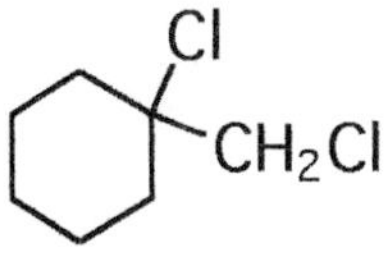

12.11 A saturated hydrocarbon contains only carbon-carbon single bonds. An unsaturated hydrocarbon contains one or more carbon-carbon double or triple bonds.

12.13 (a)

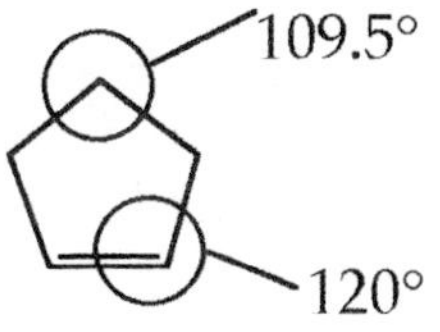

(b)

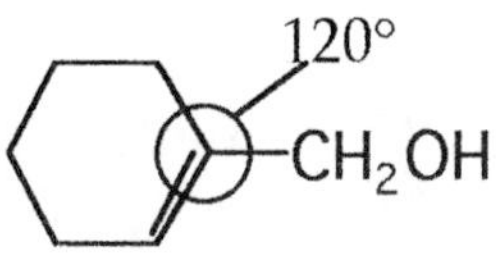

(c) 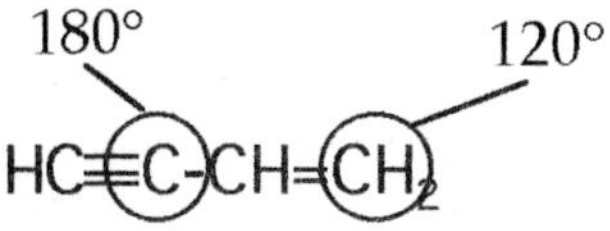 $HC{\equiv}C\text{-}CH{=}CH_2$ 180° 120°

(d)

69

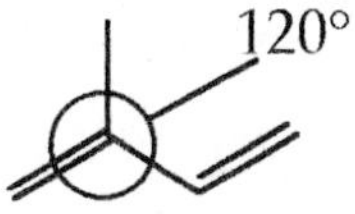

12.15 Line-angle formulas for each compound are

(a)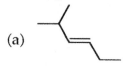

(b)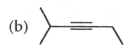

(c)

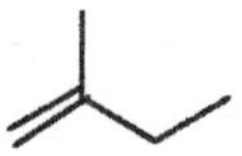

(d)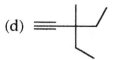

(e)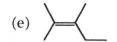

(f)

12.17 (a) 1-heptene   (b) 1,4,4-trimethylcyclopentene
     (c) 1,3-dimethylcyclohexene   (d) 2,4-dimethyl-2-pentene
     (e) 1-octyne   (f) 2,2-dimethyl-3-hexyne

12.19 (a) The longest chain is four carbons. The correct name is 2-butene.
     (b) The chain is not numbered correctly. The correct name is 2-pentene.
     (c) The ring is not numbered correctly. The correct name is 1-methylcyclohexene.
     (d) The double bond must be located. The correct name is 3,3-dimethyl-1-pentene.
     (e) The chain is not numbered correctly. The correct name is 2-hexyne.
     (f) The longest chain is five carbon atoms. The correct name is 3,4-dimethyl-2-pentene.

12.21 Parts (b), (c), and (e) show cis-trans isomerism. Following are line-angle formulas for the cis and trans isomers of each.

(b)

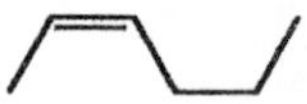

cis-2-Hexene 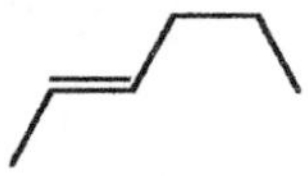   *trans*-2-Hexene

(c)

cis-3-Hexene   *trans*-3-Hexene

(e)

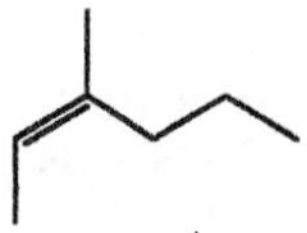

cis-3-Methyl-
2-hexene 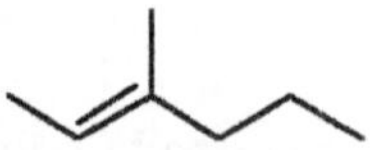   *trans*-3-Methyl-
2-hexene

12.23 There are six alkenes with the molecular formula C_5H_{10}.

1-Pentene 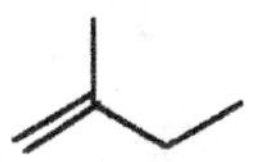   2-Methyl-1-butene 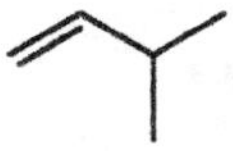   3-Methyl-1-butene

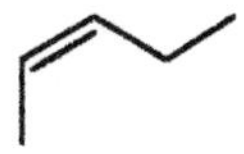

cis-2-Pentene    trans-2-Pentene 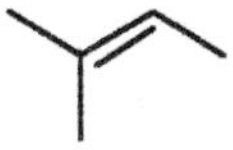   2-Methyl-2-butene

<u>12.25</u> Line-angle formulas for these three unsaturated fatty acids are

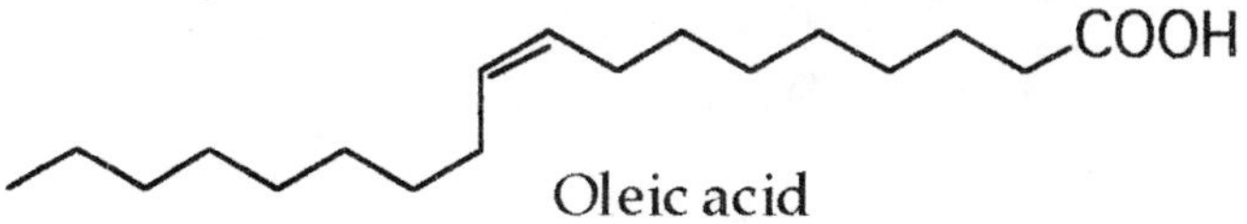

_COOH

Oleic acid

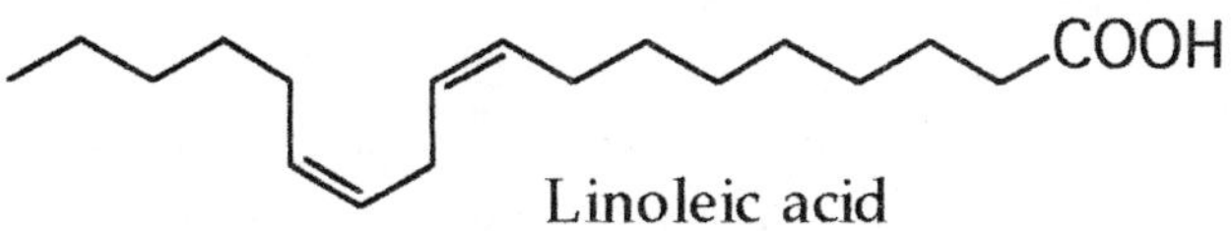

_COOH

Linoleic acid

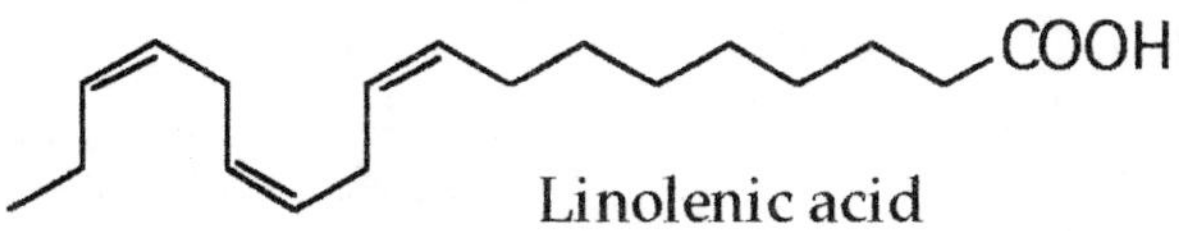

_COOH

Linolenic acid

_COOH

<u>12.27</u>

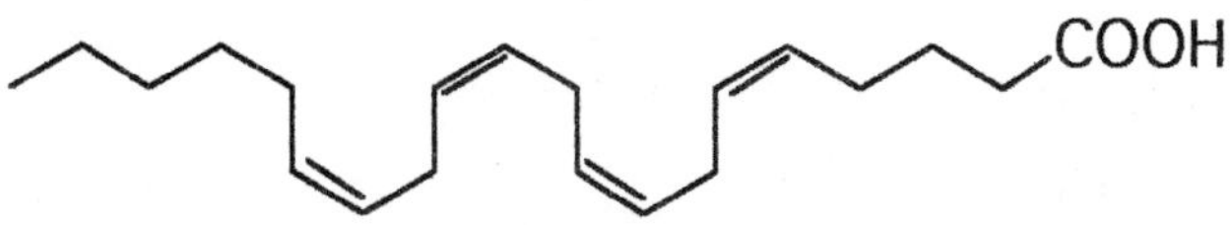

Arachidonic acid

<u>12.29</u> Only compounds (b) and (d) show cis-trans isomerism.

(b)   (d)

<u>12.31</u> The structural formula of β-ocimene is drawn on the left showing all atoms and on the right as a line-angle formula.

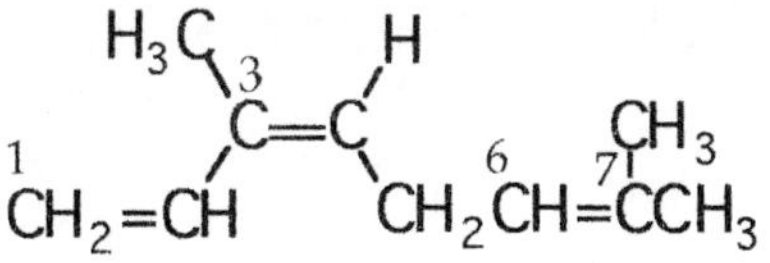

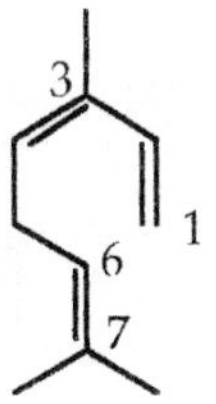

<u>12.33</u> The four isoprene units in vitamin A are shown in bold.

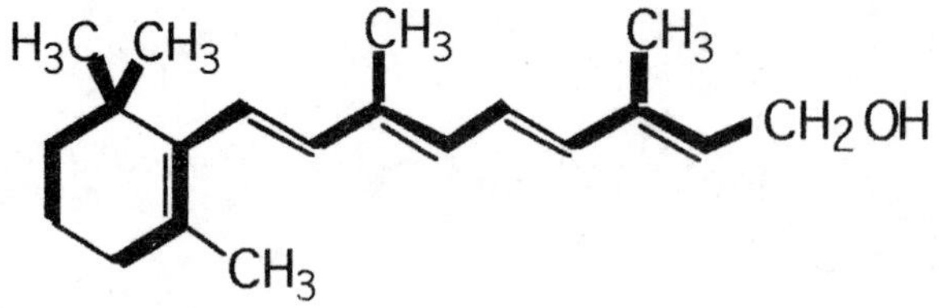

12.35 (a) HBr   (b) H_2O/H_2SO_4 (c) HI   (d) Br_2

12.37 (a) $CH_3CH_2\overset{CH_3}{\underset{+}{C}}-CH_2CH_3$ and $CH_3CH_2\overset{CH_3}{CH}\underset{+}{-CHCH_3}$

         Tertiary   Secondary

(b) $CH_3CH_2\overset{+}{CH}-CH_2CH_3$ and $CH_3CH_2CH_2-\overset{+}{C}HCH_3$

       Secondary   Secondary

(c) 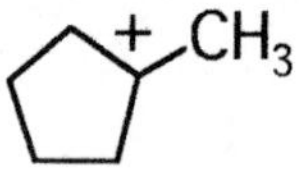   and 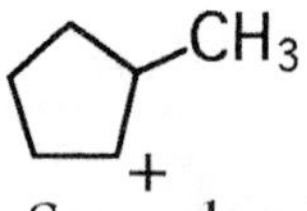   (d) 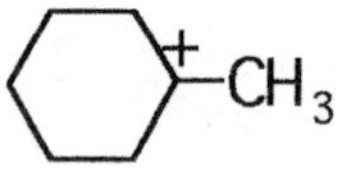   and

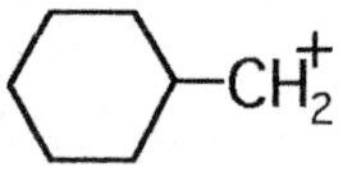

     Tertiary   Secondary   Tertiary   Primary

12.39 (a) 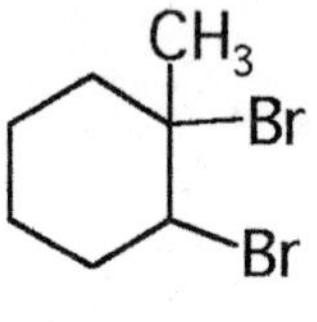   (b)

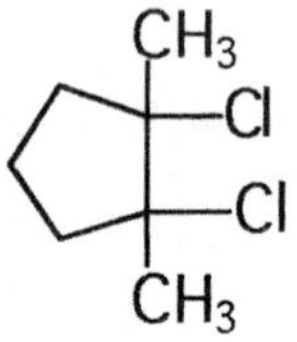

12.41 (a) $CH_3\overset{CH_3}{C}=CHCH_3$ (b) $CH_2=\overset{CH_3}{C}CH_2CH_3$ (c) $CH_2=CHCH_2CH_2CH_3$

12.43 (a) $CH_3CH_2CH=CHCH_2CH_3$ (b) 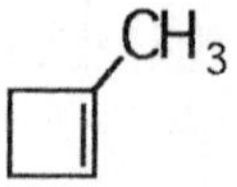 or

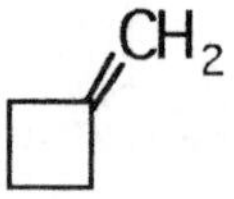

(c) $CH_2=\overset{CH_3}{C}CH_2CH_3$ or $CH_3\overset{CH_3}{C}=CHCH_3$ (d) $CH_3CH=CH_2$

12.45 (a) $CH_3CH_2CH_2CH_2CH_3$ (b) $CH_3CH_2CH_2CH_2CH_3$

(c)    (d)

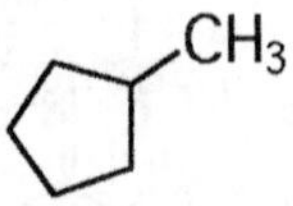

12.47 Reagents are shown over each arrow.

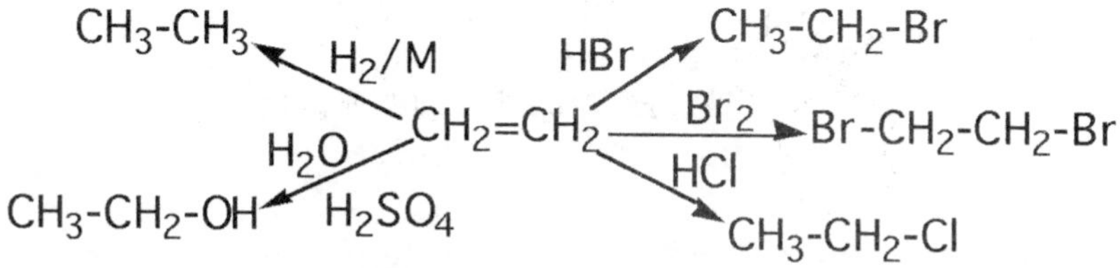

$$CH_3-CH_3 \xleftarrow{H_2/M} CH_2=CH_2 \xrightarrow{HBr} CH_3-CH_2-Br$$

$$CH_3-CH_2-OH \xleftarrow[H_2SO_4]{H_2O} CH_2=CH_2 \xrightarrow[HCl]{Br_2} Br-CH_2-CH_2-Br$$

$$CH_3-CH_2-Cl$$

12.49 Ethylene is a natural ripening agent for fruits.

12.51 Its molecular formula is $C_{16}H_{30}O_2$ and its molecular weight is 254.4 amu, and molar mass is 254.4 g/ mol.

12.53 Rods are primarily responsible for black-and-white vision. Cones are responsible for color vision.

12.55 The most common consumer items made of high-density polyethylene (HDPE) are milk and water jugs, grocery bags, and squeeze bottles. The most common consumer items made of low-density polyethylene (LDPE) are packaging for baked goods, vegetables and other produce, as well as trash bags. Currently only HDPE materials are recyclable.

12.57 There are five compounds with the molecular formula C_4H_8. All are constitutional isomers. The only cis-trans isomers are *cis*-2-butene and *trans*-2-butene.

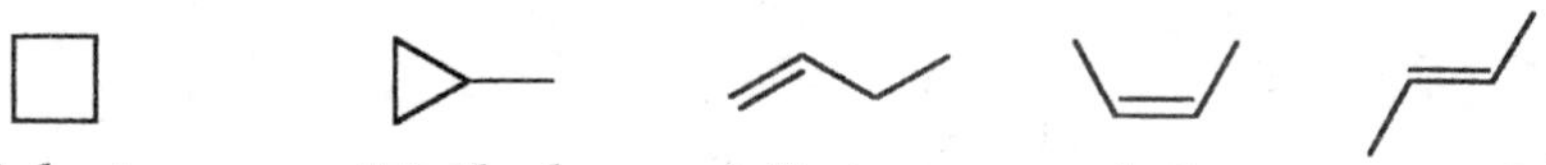

Cyclobutane	Methyl-cyclopropane	1-Butene	*cis*-2-Butene	*trans*-2-Butene

12.59 (a) There are five alkenes with this carbon skeleton.

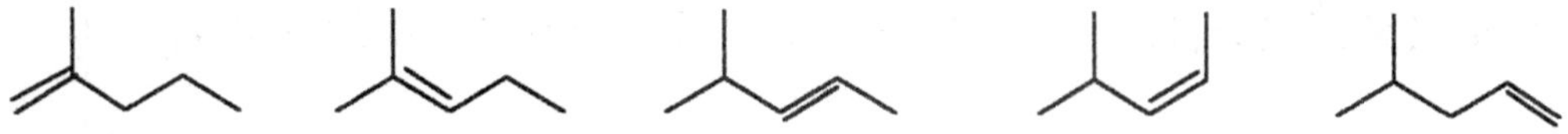

2-Methyl-1-pentene	2-Methyl-2-pentene	*trans*-4-Methyl-2-pentene	*cis*-4-Methyl-2-pentene	4-Methyl-1-pentene

(b) There are two alkenes with this carbon skeleton.

2,3-Dimethyl-1-butene	2,3-Dimethyl-2-butene

(c) There is one alkene with this carbon skeleton.

73

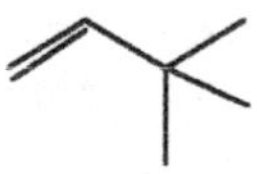

3,3-Dimethyl-1-butene

12.61 The carbon skeletons of lycopene and β-carotene are almost identical. The difference is that, at each end, the second carbon of β-carotene is bonded to the seventh carbon to form the two six-membered rings of lycopene.

12.63 Each alkene hydration reaction follows Markovnikov's rule. In (a), -H adds preferentially to carbon-1 and -OH to carbon-2 to give 2-hexanol. In (b), each carbon of the double bond has the same pattern of substitution, so 2-hexanol and 3-hexanol are formed in approximately equal amounts. In (c), each carbon of the double bond again has the same pattern of substitution, but no matter which way H-OH adds, the only product possible is 3-hexanol.

(a) 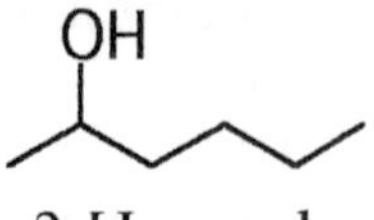 2-Hexanol   (b) 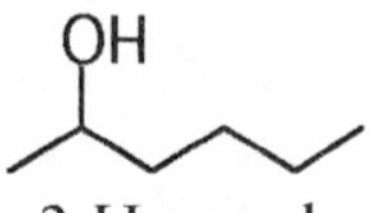 2-Hexanol + 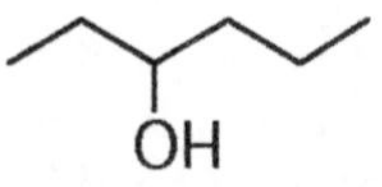 3-Hexanol   (c) 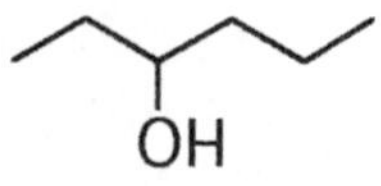 3-Hexanol

12.65 (a)    (b) 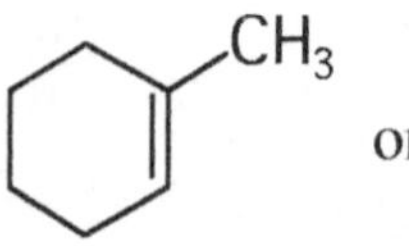   or 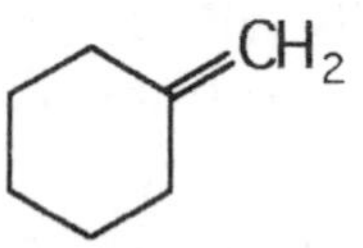   (c)

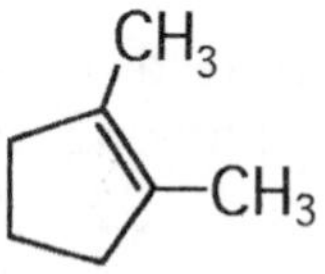

(d) 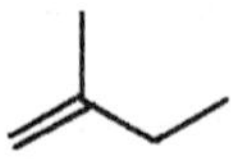   or

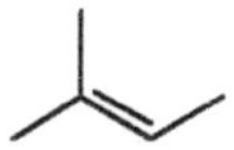

Chapter 13 Benzene and its Derivatives

13.1 (a) 2,4,6-tri-*tert*-butylphenol   (b) 2,4-dichloroaniline
       (c) 3-nitrobenzoic acid

13.3 An aromatic compound is one that contains one or more benzene rings.

13.5 Yes, they have double bonds, at least in the contributing structures we normally use to represent them. Yes, they are unsaturated because they have fewer hydrogens than a cycloalkane with the same number of carbons.

13.7 (a) An alkene of six carbons has the molecular formula C_6H_{12} and contains one carbon-carbon double bond. Three examples are

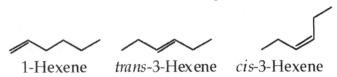

1-Hexene   *trans*-3-Hexene   *cis*-3-Hexene

(b) A cycloalkene of six carbons has the molecular formula C_6H_{10} and contains one ring and one carbon-carbon double bond. Three examples are

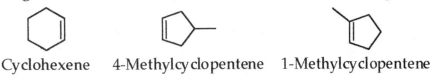

Cyclohexene   4-Methylcyclopentene   1-Methylcyclopentene

(c) An alkyne of six carbons has the molecular formula C_6H_{10} and contains one carbon-carbon triple bond. Three examples are

1-Hexyne   2-Hexyne   4-Methyl-2-pentyne

(d) An aromatic hydrocarbon of eight carbons has the molecular formula C_8H_{10} and contains one benzene ring. Three examples are

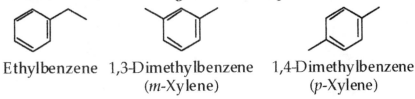

Ethylbenzene 1,3-Dimethylbenzene   1,4-Dimethylbenzene
              (*m*-Xylene)   (*p*-Xylene)

13.9 Benzene consists of carbons, each surrounded by three regions of electron density, which gives 120° for all bond angles. Bond angles of 120° in benzene can be maintained only if the molecule is planar. Cyclohexane, on the other hand, consists of carbons, each surrounded by four regions of electron density, which gives 109.5° for all bond angles. Angles of 109.5° in cyclohexane can be maintained only if the molecule is nonplanar.

13.11 It works this way. Neither a unicorn, which has a horn like a rhinoceros, nor a dragon, which has a tough, leathery hide like a rhinoceros, exists. If they did and you made a hybrid of them, you would have a rhinoceros. Furthermore, a rhinoceros is not a dragon part of the time and a unicorn the rest of the time; a rhinoceros is a rhinoceros all of the time. To carry this analogy to aromatic compounds, resonance-contributing structures for them do not exist; they are imaginary. If they did exist and you could make a hybrid of them, you would have the real aromatic compound.

13.13 (a) 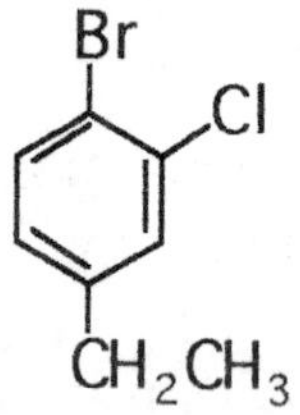 (b) 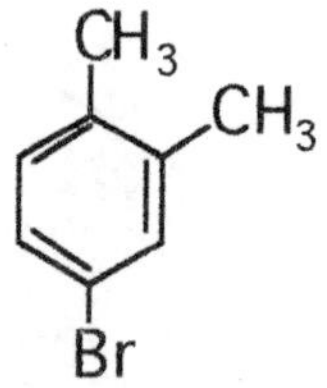 (c)

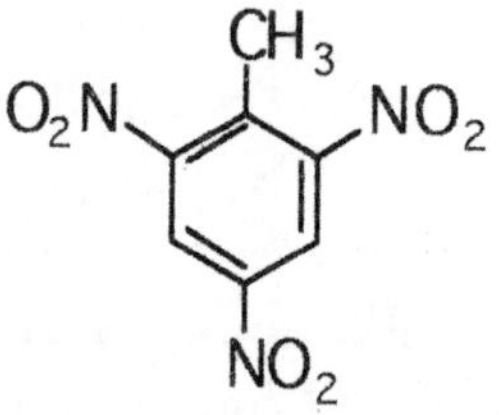

(d) 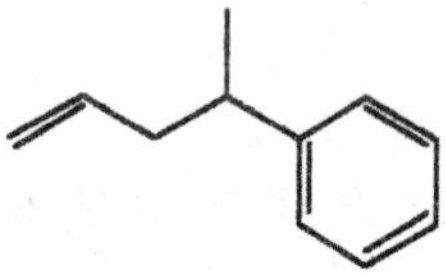 (e) 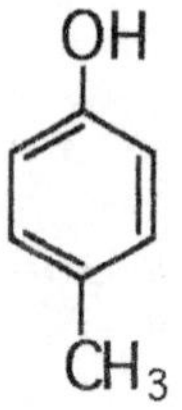 (f)

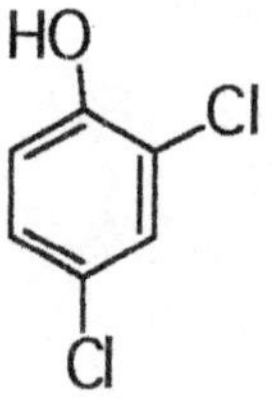

13.15 Only cyclohexene will react with a solution of bromine in dichloromethane. A solution of Br_2/CH_2Cl_2 is red, whereas a dibromocycloalkane is colorless. To tell which bottle contains which compound, place a small quantity of each compound in a test tube and to each add a few drops of Br_2/CH_2Cl_2 solution. If the red color disappears, the compound is cyclohexene. If it remains, the compound is benzene.

13.17

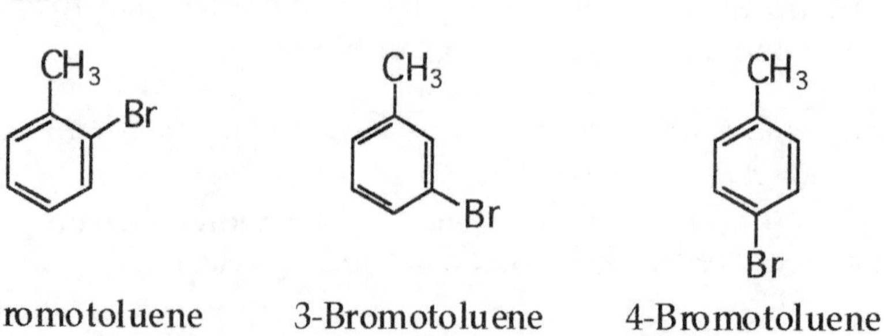

2-Bromotoluene
(o-Bromotoluene)

3-Bromotoluene
(m-Bromotoluene)

4-Bromotoluene
(p-Bromotoluene)

13.19 (a) nitration using HNO_3/H_2SO_4 followed by sulfonation using H_2SO_4. The order of the steps may also be reversed.

(b) bromination using $Br_2/FeCl_3$ followed by chlorination using $Cl_2/FeCl_3$. The order of the steps may also be reversed.

13.21 Phenol is a sufficiently strong acid that it reacts with strong bases such as sodium hydroxide to form sodium phenoxide, a water-soluble salt. Cyclohexanol has no comparable acidity and does not react with sodium hydroxide.

13.23 *Radical* indicates a molecule or ion with an unpaired electron. *Chain* means a cycle of two or more steps, called propagation steps, that repeat over and over. The net effect of a radical chain reaction is the conversion of starting materials to products. *Chain length* refers to the number of times the cycle of chain propagation steps repeats.

13.25 Vitamin E participates in one or the other of the chain propagation steps and forms a stable radical, which breaks the cycle of propagation steps.

13.27 The abbreviation DDT is derived from DichloroDiphenylTrichloroethane.

13.29 Insoluble in water. It is entirely a hydrophobic molecule.

13.31 Biodegradable means that a substance can be broken down into one or more harmless compounds by living organisms in the environment.

13.33 Iodine is an element that is found primarily in seawater and, therefore, seafoods are rich sources of it. Individuals in inland areas where seafood is only a limited part of the diet are the most susceptible to developing goiter.

13.35 One of the aromatic rings in Allura red has a -CH$_3$ group and an -OCH$_3$ group. These groups are not present in Sunset Yellow.

13.37 orange

13.39 Capsaicin is isolated from the fruit of various species of *Capsicum*, otherwise known as chili peppers.

13.41 Two cis-trans isomers are possible for capsaicin.

13.43 Following are the three contributing structures for naphthalene.

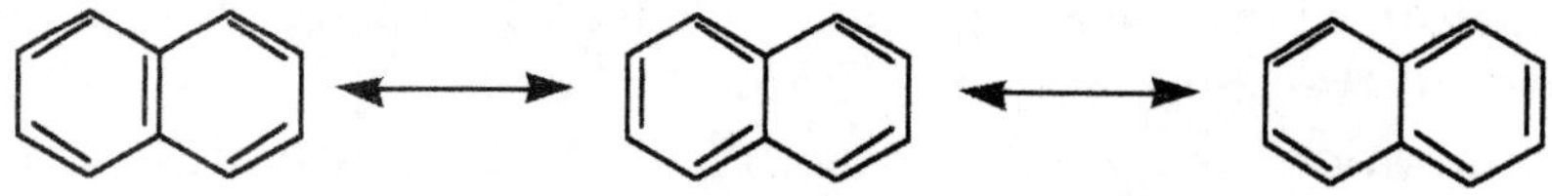

(1)   (2)   (3)

<u>13.45</u> BHT participates in one of the chain propagation steps of autoxidation, forms
a stable radical, and thus terminates autoxidation.

<u>13.47</u>

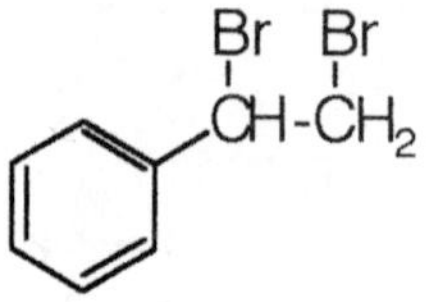

Chapter 14 Alcohols, Ethers, and Thiols

14.1 (a) 2-heptanol   (b) 2,2-dimethyl-1-propanol
(c) *trans*-3-isopropylcyclohexanol

14.2 (a) primary   (b) secondary   (c) primary   (d) tertiary

14.3 In each case, the major product (circled) contains the more substituted double bond.

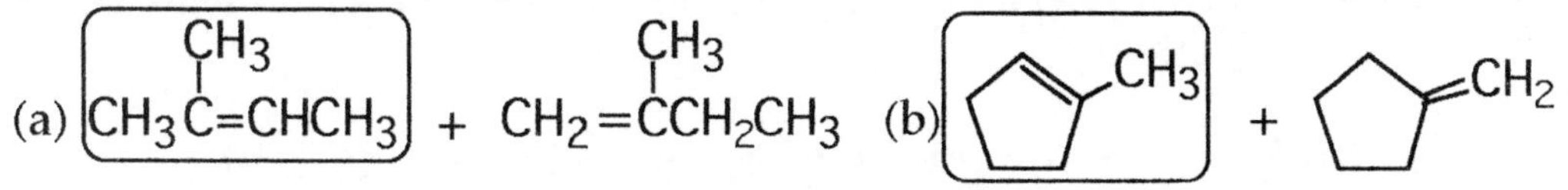

$$(a) \boxed{CH_3\ \underset{\underset{CH_3}{|}}{C}=CHCH_3} + CH_2=\underset{\underset{CH_3}{|}}{C}CH_2CH_3 \quad (b)$$

14.4

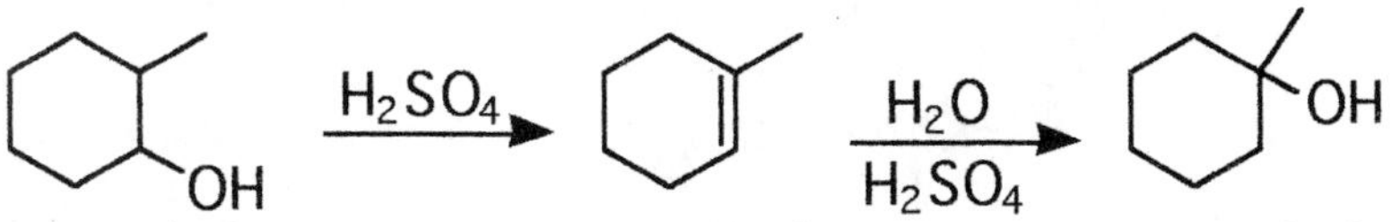

2-Methyl-   1-Methyl-   1-Methyl-
cyclohexanol   cyclohexene (C)   cyclohexanol (D)

14.5 Each secondary alcohol is oxidized to a ketone.

(a)

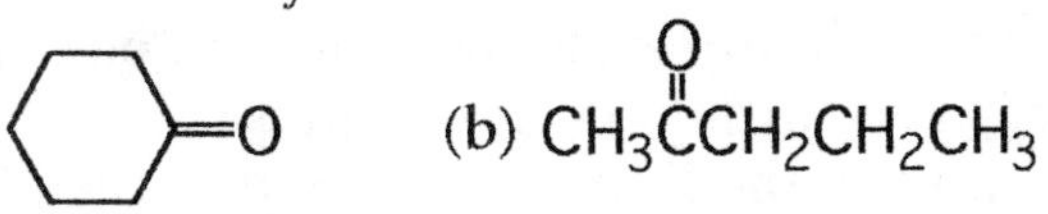

(b) $CH_3\overset{\overset{O}{||}}{C}CH_2CH_2CH_3$

14.6 (a) ethyl isobutyl ether   (b) cyclopentyl methyl ether

14.7 (a) 3-methyl-1-butanethiol   (b) 3-methyl-2-butanethiol

14.9 Only (c) and (d) are secondary alcohols.

14.11 (a) 1-pentanol   (b) 1,3-propanediol
(c) 1,2-butanediol   (d) 3-methyl-1-butanol
(e) *cis*-1,2-cyclohexanediol   (f) 2,6-dimethylcyclohexanol

14.13 (a)

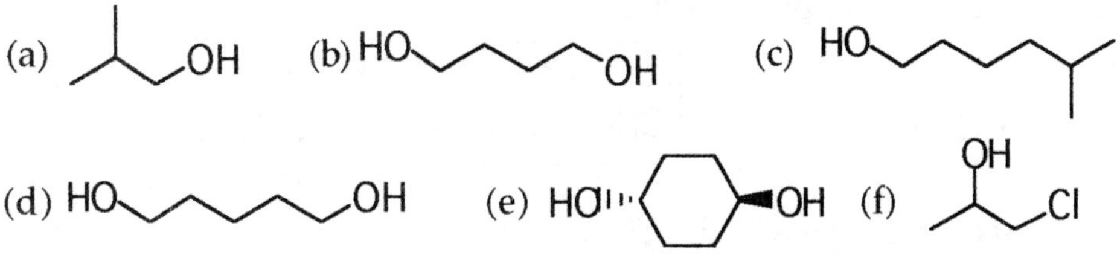

14.15 (a) ethyl isopropyl ether   (b) dibutyl ether   (c) diphenyl ether

14.17 (a) *sec*-butyl mercaptan   (b) butyl mercaptan   (c) cyclohexyl mercaptan

14.19 Prednisone contains three ketones, one primary alcohol, one tertiary alcohol, one disubstituted carbon-carbon double bond, and one trisubstituted carbon-carbon double bond. Estradiol contains one secondary alcohol and one disubstituted phenol.

14.21 Low-molecular-weight alcohols form hydrogen bonds with water molecules through both the oxygen and hydrogen atoms of their -OH groups. Low-molecular-weight ethers form hydrogen bonds with water molecules only through the oxygen atom of their -O- groups. The greater extent of hydrogen bonding between alcohol and water molecules makes the low-molecular-weight alcohols more soluble in water than the low-molecular-weight ethers.

14.23 Both types of hydrogen bonding are shown on the following illustration.

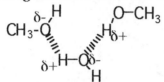

14.25 In order of increasing boiling point, they are

$CH_3CH_2CH_3$	CH_3CH_2OH	$CH_3CH_2CH_2CH_2OH$	$HOCH_2CH_2OH$
-42°C	78°C	117°C	198°C

14.27 Because 1-butanol molecules associate in the liquid state by hydrogen bonding, it has the higher boiling point (117°C). There is very little polarity to an S-H bond. The only interactions among 1-butanethiol molecules in the liquid state are the considerably weaker London dispersion forces. For this reason, 1-butanethiol has the lower boiling point (98°C).

14.29 Evaporation of a liquid from the surface of the skin cools because heat is absorbed from the skin in converting molecules from the liquid state to the gaseous state. 2-Propanol (isopropyl alcohol), which has a boiling point of 82°C, absorbs heat from the skin, evaporates rapidly, and has a cooling effect. 2-Hexanol, which has a boiling point of 140°C, also absorbs heat from the surface of the skin but, because of its higher boiling point, evaporates much more slowly and, therefore, does not have the same cooling effect as 2-propanol.

14.31 The more water-soluble compound is circled.

(a) $\boxed{CH_3OH}$ or CH_3OCH_3 (b) $\boxed{CH_3\overset{OH}{\underset{|}{C}HCH_3}}$ or $CH_3\overset{CH_2}{\underset{||}{C}CH_3}$

(c) $CH_3CH_2CH_2SH$ or $\boxed{CH_3CH_2CH_2OH}$

14.33 For three parts, two constitutional isomers will give the desired alcohol. For two parts, only one alkene will give the desired alcohol.

(a) $CH_2=CHCH_2CH_3$ or $CH_3CH=CHCH_3$ (b) [structure: cyclohexene with CH_3] or [structure: methylenecyclohexane with CH_2]

(c) $CH_3CH_2CH=CHCH_2CH_3$ (d) $CH_2=\overset{CH_3}{\underset{}{C}}CH_2CH_2CH_3$ or $CH_3\overset{CH_3}{\underset{}{C}}=CHCH_2CH_3$

(e) [structure: cyclopentane pentagon]

14.35 Phenols are weak acids, with pK_a values approximately equal to 10. Alcohols, considerably weaker acids, have about the same acidity as water.

14.37 The first reaction is an acid-catalyzed dehydration; the second is an oxidation.

(a) $CH_3CH_2CH_2CH_2OH \xrightarrow[\text{heat}]{H_2SO_4} CH_3CH_2CH=CH_2 + H_2O$

(b) $CH_3CH_2CH_2CH_2OH \xrightarrow[H_2SO_4]{K_2Cr_2O_7} CH_3CH_2CH_2\overset{O}{\overset{\|}{C}}OH$

14.39 Oxidation of a primary alcohol by $K_2Cr_2O_7/H_2SO_4$ gives a carboxylic acid.

(a) $CH_3(CH_2)_6\overset{O}{\overset{\|}{C}}OH$ (b) $HO\overset{O}{\overset{\|}{C}}CH_2CH_2\overset{O}{\overset{\|}{C}}OH$

14.41 Each can be prepared from 1-propanol (circled) as shown in this flow chart.

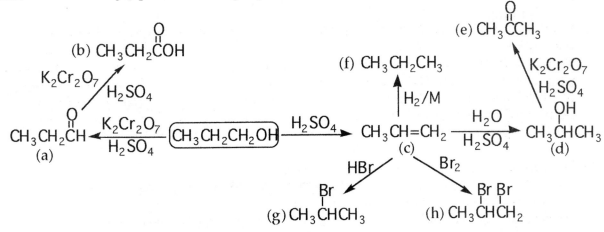

14.43 Ethanol and ethylene glycol are derived from ethylene. Ethanol is a solvent and is the starting material for the synthesis of diethyl ether, also an important solvent. Ethylene glycol is used in automotive antifreezes and is one of the two starting materials required for the synthesis of poly(ethylene terephthalate), better known as PET (Section 18.8B).

14.45 (a) The three functional groups are a thiol, a primary amine, and a carboxylic acid. (b) Oxidation of the thiol group gives a disulfide (-S-S-).

81

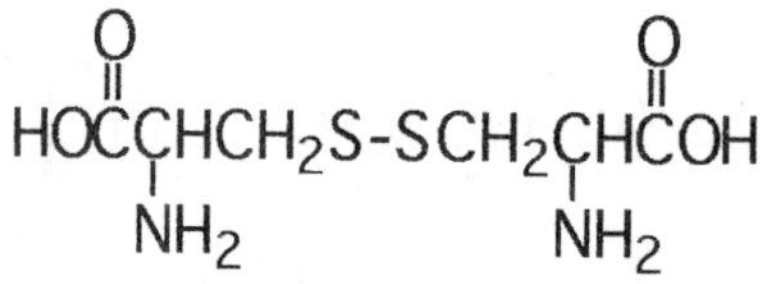

$$\underset{\underset{NH_2}{|}}{HOC}CHCH_2S\text{-}SCH_2\underset{\underset{NH_2}{|}}{C}HCOH$$

with carbonyl oxygens on both terminal carbons.

14.47 Nitroglycerin was discovered in 1847. It is a pale yellow, oily liquid.

14.49 One of the products the body derives from metabolism of nitroglycerin is nitric oxide, NO, which causes the coronary artery to dilate, thus relieving angina.

14.51 The relationship is that 2100 mL of breath contains the same amount of ethanol as 1.00 mL of blood.

14.53 Diethyl ether is easy to use and causes excellent muscle relaxation. Blood pressure, pulse rate, and respiration are usually only slightly affected. Diethyl ether's chief drawbacks are its irritating effect on the respiratory passages and its after-effect of nausea.

14.55 Enflurane and isoflurane are insoluble in water but soluble in hexane.

14.57 $2CH_3OH + 3O_2 \longrightarrow 2CO_2 + 4H_2O$

14.59 The eight isomeric alcohols with the molecular formula $C_5H_{12}O$ are

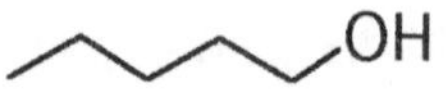

1-Pentanol 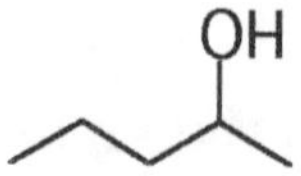   2-Pentanol 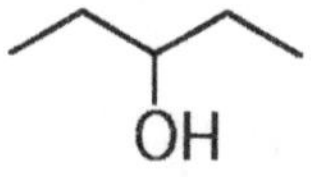   3-Pentanol 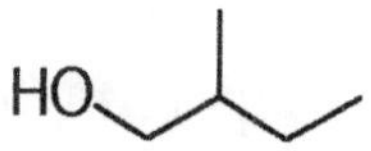   2-Methyl-1-butanol

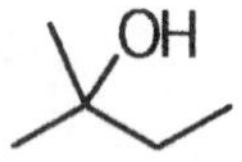

2-Methyl-2-butanol 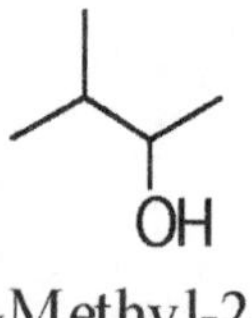   3-Methyl-2-butanol 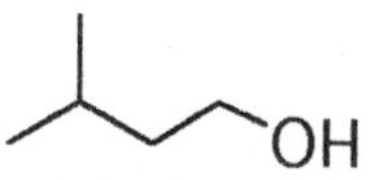   3-Methyl-1-butanol 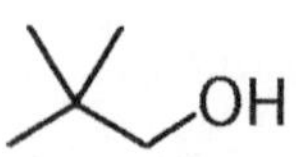   2,2-Dimethyl-1-propanol

14.61 Ethylene glycol has two -OH groups by which each molecule participates in hydrogen bonding, whereas 1-propanol has only one. The stronger intermolecular forces of attraction between molecules of ethylene glycol give it the higher boiling point.

14.63 Arranged in order of increasing solubility in water, they are
$CH_3CH_2CH_2CH_2CH_2CH_3$ $CH_3CH_2CH_2CH_2CH_2OH$ $HOCH_2CH_2CH_2CH_2OH$
Hexane (insoluble)   1-Pentanol (2.3 g/mL water)   1,4-Butanediol (infinitely soluble)

14.65 Each can be prepared from 2-methyl-1-propanol (circled) as shown in this flow chart.

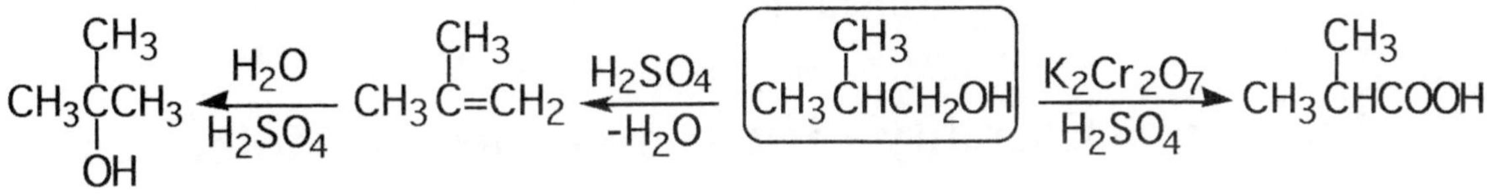

$$CH_3\underset{OH}{\overset{CH_3}{\underset{|}{\overset{|}{C}}}}CH_3 \xleftarrow[H_2SO_4]{H_2O} CH_3\overset{CH_3}{\overset{|}{C}}{=}CH_2 \xleftarrow[-H_2O]{H_2SO_4} \boxed{CH_3\overset{CH_3}{\overset{|}{C}}HCH_2OH} \xrightarrow[H_2SO_4]{K_2Cr_2O_7} CH_3\overset{CH_3}{\overset{|}{C}}HCOOH$$

Chapter 15 Chirality - The Handedness of Molecules

15.1 The enantiomers of each part are drawn with two groups in the plane of the paper, a third group toward you in front of the plane, and the fourth group away from you behind the plane.

(a)

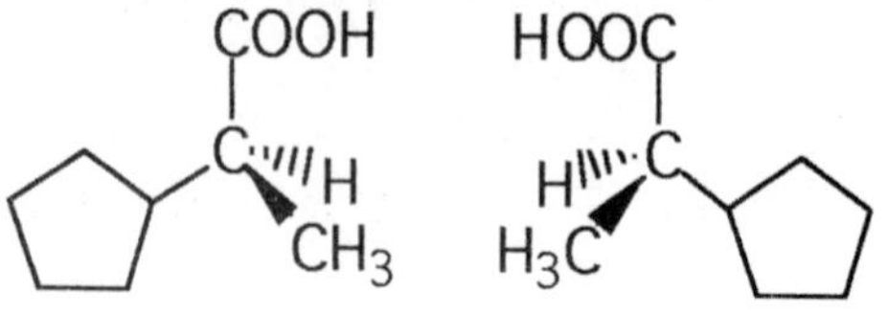

(b)

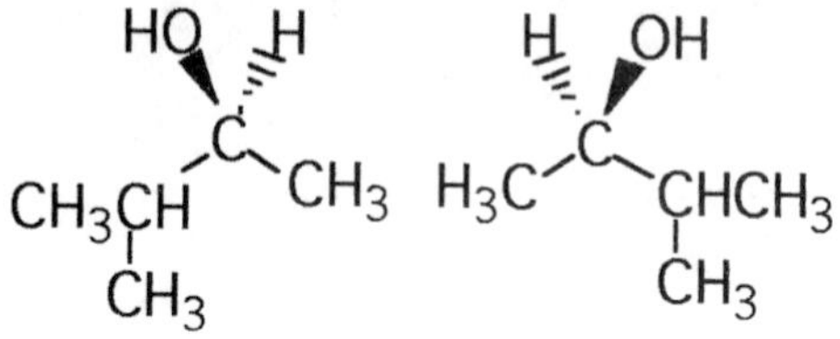

15.2 The group of higher priority in each set is circled.

(a) $\boxed{-CH_2OH}$ and $-CH_2CH_2COOH$ (b) $\boxed{-CH_2NH_2}$ and $-CH_2CH_2COOH$

15.3 The order of priorities and the configuration of each are shown in the drawings.

(a)   (b)

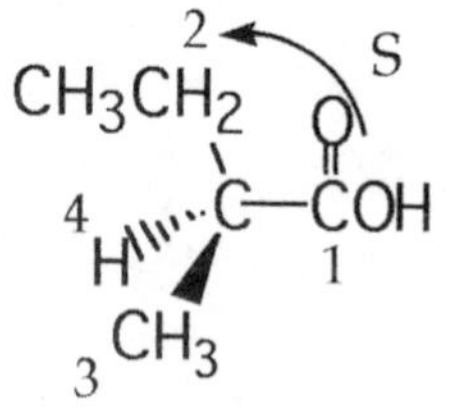

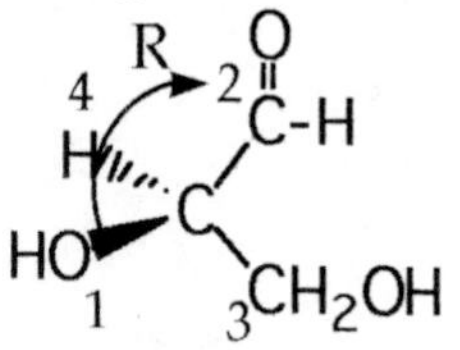

15.4 (a) Compounds 1 and 3 are one pair of enantiomers. Compounds 2 and 4 are a second pair of enantiomers. (b) Compounds 1 and 2, 1 and 4, 2 and 3, and 3 and 4 are diastereomers.

15.5 Four stereoisomers are possible for 3-methylcyclohexanol. The cis isomer is one pair of enantiomers; the trans isomer is a second pair of enantiomers.

15.6 Stereocenters are marked by an asterisk and the number of stereoisomers possible is shown under the structural formula.

(a)

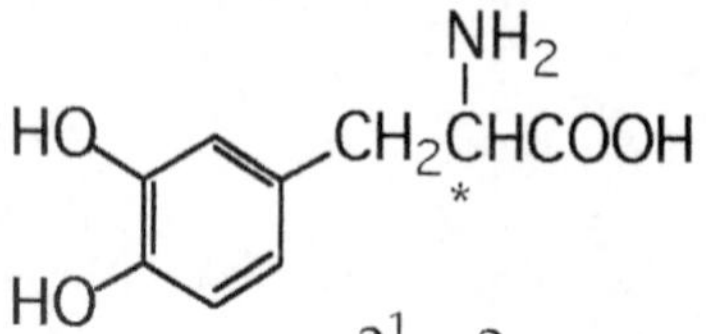

$2^1 = 2$

(b) $CH_2=CHCHCH_2CH_3$ with OH on the starred carbon

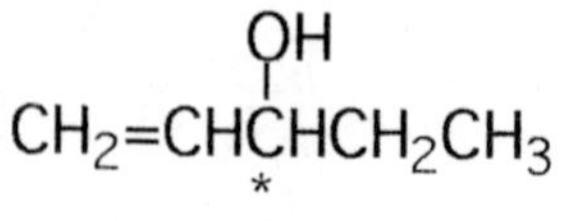

$2^1 = 2$

(c)

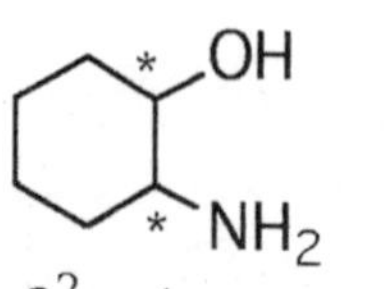

$2^2 = 4$

15.7 Chirality is a property of an object. The object is not superposable on its mirror image; that is, the object has handedness. 2-Butanol is a chiral molecule.

15.9 Stereoisomers are isomers that have the same molecular formula and the same connectivity, but a different orientation of their atoms in space. Three examples are cis-trans isomers, enantiomers, and diastereomers.

Chapter 15 Chirality

15.11 (a) Chiral.   (b) Achiral.   (c) Achiral.   (d) Achiral.
(e) Chiral, unless you are on the equator, in which case it goes straight down and has no chirality.

15.13 Neither *cis*-2-butene nor *trans*-2-butene is chiral. Each is superposable on its mirror image.

15.15 Compounds (a), (c), and (d) contain stereocenters, here marked by asterisks.

(a) 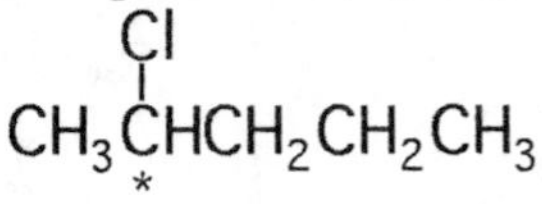 $CH_3\overset{Cl}{\underset{*}{C}}HCH_2CH_2CH_3$ (c) $CH_2=CH\overset{Cl}{\underset{*}{C}}HCH_3$ (d) 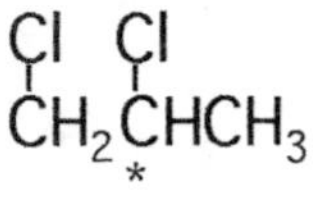 $CH_2\overset{Cl\ Cl}{\underset{*}{C}}HCH_3$

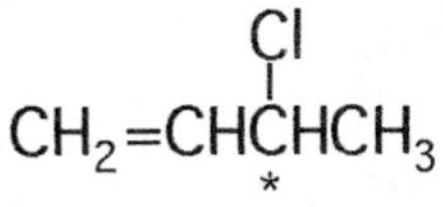

15.17 The stereocenter in each chiral compound is marked by an asterisk.

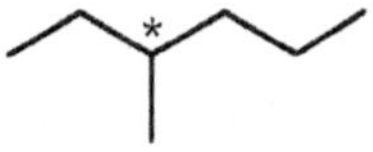

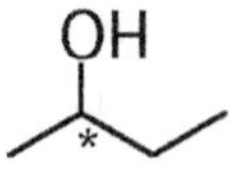

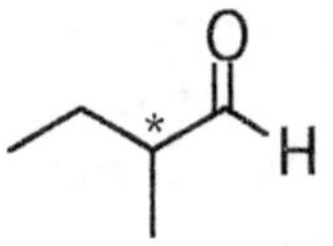

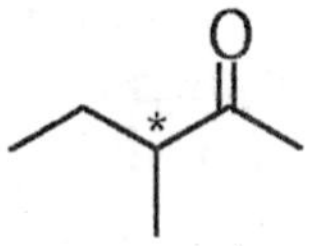

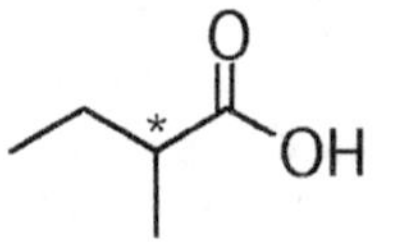

15.19 Following are mirror images of each.

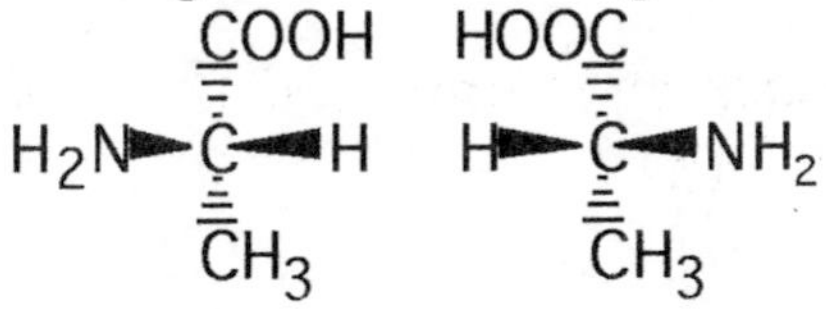

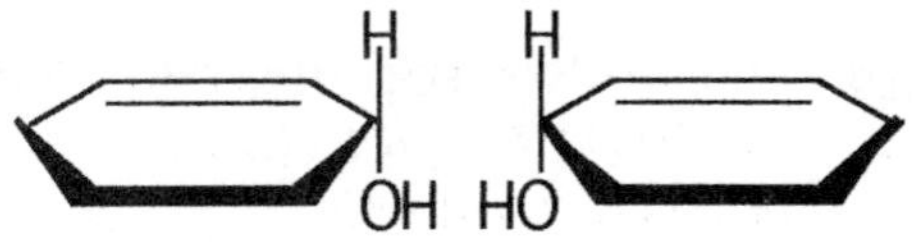

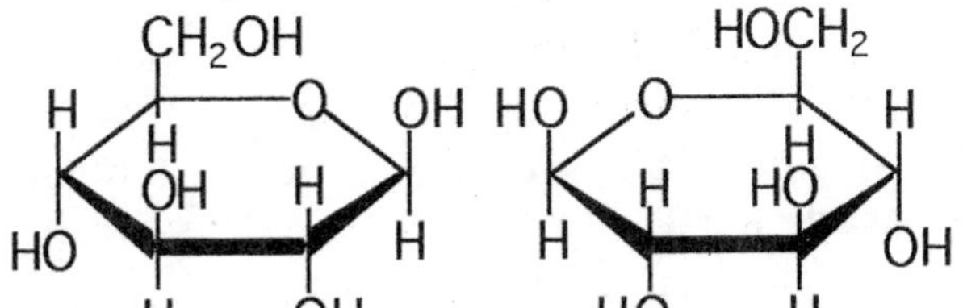

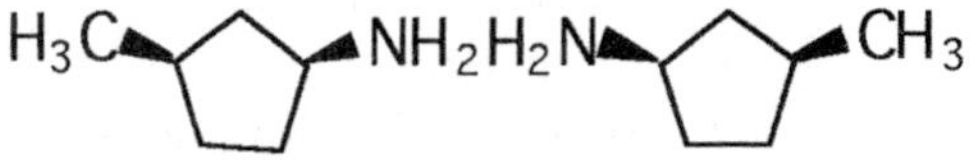

15.21 Parts (b) and (c) contain stereocenters.

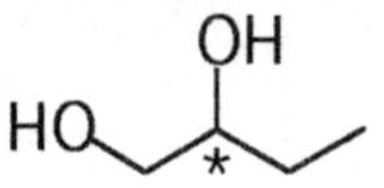

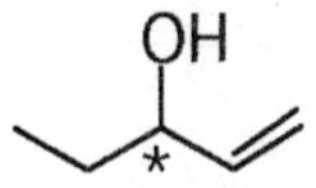

15.23 Each stereocenter is marked by an asterisk. Under each structural formula is the number of stereoisomers possible. Compound (b) has no stereocenter.

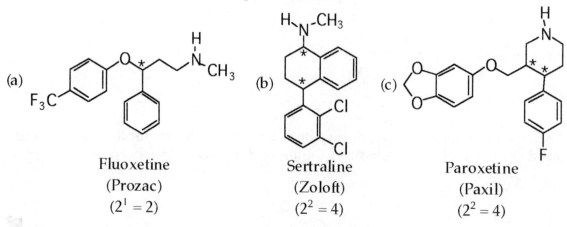

(a) $\overset{OH}{\text{...}}$ (c) ...OH (d) ...

$(2^2 = 4)$ $(2^1 = 2)$ $(2^2 = 4)$

<u>15.25</u> Molecules (b) and (d) have R configurations. Molecules (a) and (c) have S configurations.

(a) CH_3 ... Br ... S CH_2OH (b) CH_3 ... $HOCH_2$ R Br (c) $\overset{2}{C}H_2OH$ ... H_3C ... Br (d) R $\overset{2}{C}H_2OH$ ... Br CH_3

<u>15.27</u> The optical rotation of its enantiomer is +41°.

<u>15.29</u> To say that a drug is chiral means that it has handedness - that it has one or more stereocenters and the possibility for two or more stereoisomers. Just because a compound is chiral does not mean that it will be optically active. It may be chiral and present as a racemic mixture, in which case it will have no effect on the plane of polarized light. If, however, it is present as a single enantiomer, it will rotate the plane of polarized light.

<u>15.31</u> The two stereocenters are marked by asterisks.

3-Methyl-2-pentanol $\overset{OH}{\text{* *}}$

<u>15.33</u> Each stereocenter is marked by an asterisk. Under the name of each compound is the number of stereoisomers possible for it.

(a) Fluoxetine (Prozac) $(2^1 = 2)$

(b) Sertraline (Zoloft) $(2^2 = 4)$

(c) Paroxetine (Paxil) $(2^2 = 4)$

86

15.35 (a) In this chair conformation of glucose, carbons 1, 2, 3, 4, and 5 are stereocenters. (b) There are $2^5 = 32$ stereoisomers possible. (c) Because enantiomers always occur in pairs, there are 16 pairs of enantiomers possible.

15.37 (a) The eight stereocenters are marked by asterisks. (b) There are $2^8 = 256$ stereoisomers possible.

Triamcinolone acetonide

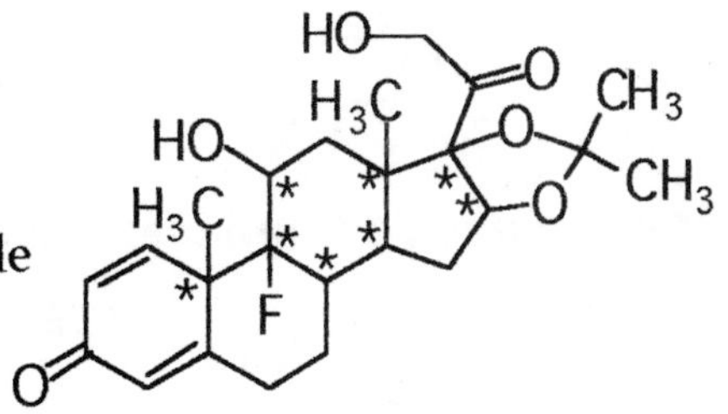

15.39 The majority have a right-handed twist because the machines that make them all impart the same twist.

Chapter 16 Amines

Chapter 16 Amines

16.1 Pyrollidine has nine hydrogens; its molecular formula is C_4H_9N. Purine has four hydrogens; its molecular formula is $C_5H_4N_4$.

16.2 Following is a line-angle formula for each compound.

(a) [line-angle structure] —NH_2 (b) [cyclopentane ring] —NH_2 (c) H_2N [chain] NH_2

16.3 Following is a line-angle formula for each compound.

(a) HO [chain] NH_2 (b) [phenyl]—$\underset{H}{N}$—[phenyl]   (c) [isopropyl]$\underset{H}{N}$[isopropyl]

16.4 The stronger base is circled.

(a) [pyridine] N or [cyclohexyl]—NH_2 (b) NH_3 or [phenyl]—NH_2

16.5 The product of each reaction is an amine salt.

(a) $(CH_3CH_2)_3\overset{+}{N}H$ Cl^- (b) [piperidine ring]$\overset{H}{\underset{H}{\overset{+}{N}}}$ $CH_3\overset{O}{\overset{\|}{C}}O^-$

16.7 In an aliphatic amine, all carbon groups bonded to nitrogen are alkyl groups. In an aromatic amine, one or more of the carbon groups bonded to nitrogen are aryl (aromatic) groups.

16.9 Following is a structural formula for each amine.

(a) [structure with NH_2]   (b) $CH_3(CH_2)_6CH_2NH_2$ (c) [structure with NH_2]

(d) $H_2N(CH_2)_5NH_2$ (e) [benzene with NH_2 and Br]   (f) $(CH_3CH_2CH_2CH_2)_3N$

16.11 Each amine is classified by type.

1° aliphatic amine

HO

heterocyclic aromatic amine

(a)

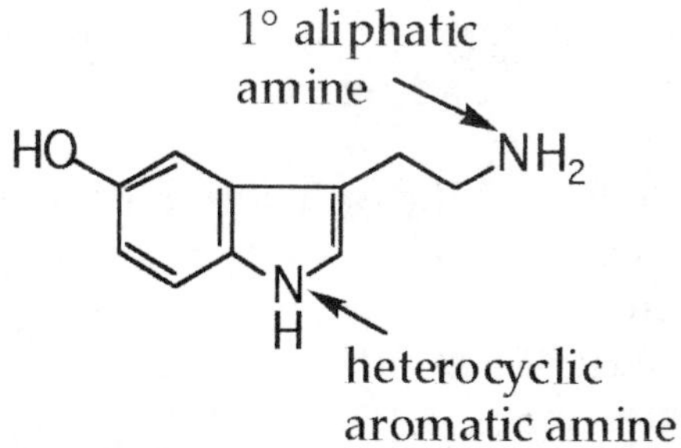

(b)

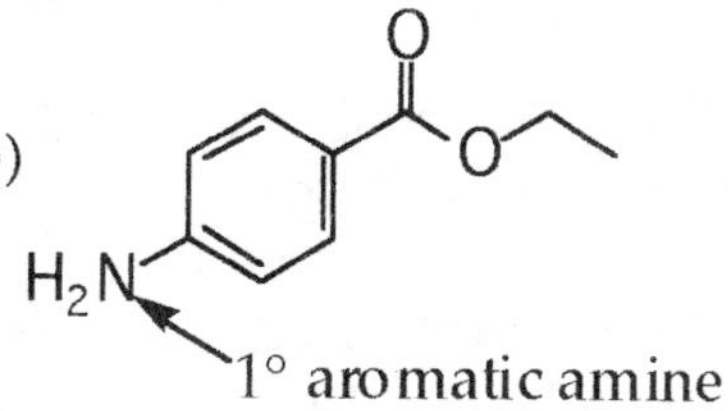

H_2N 1° aromatic amine

3° aliphatic amine

CH_3

$N-CH_3$

(c)

Dimethylethylamine

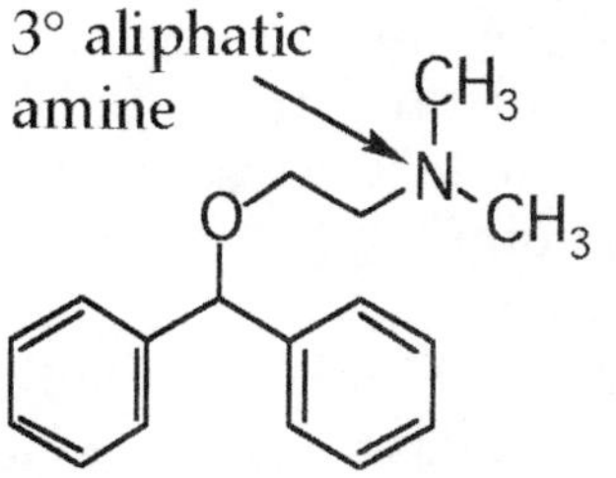

16.13 There are four primary amines of this molecular formula, three secondary amines, and one tertiary amine. Only 2-butanamine is chiral.

1° amines:

1-Butanamine (Butylamine) 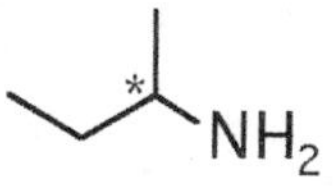 2-Butanamine (sec-Butylamine) 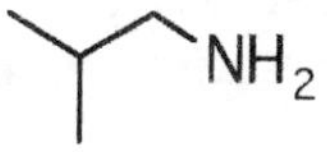 2-Methyl-1-propanamine (Isobutylamine)  1,1-Dimethyl-ethanamine (*tert*-Butylamine)

2° amines:

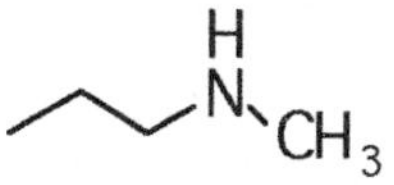

Methylpropyl-amine 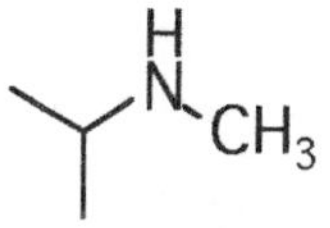 Methylisopropyl-amine 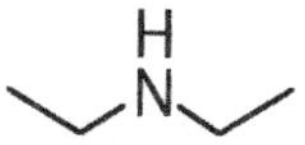 Diethylamine

3° amine:

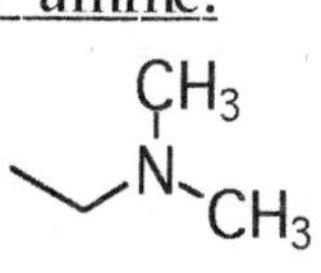

CH_3

$N-CH_3$

Dimethylethylamine

16.15 Both propylamine (a 1° amine) and ethylmethylamine (a 2° amine) have an N-H group and hydrogen bonding occurs between their molecules in the liquid state. Because of this intermolecular force of attraction, these two amines have higher boiling points than trimethylamine, which has no N-H bond and, therefore, cannot participate in intermolecular hydrogen bonding.

16.17 2-Methylpropane is a nonpolar hydrocarbon and the only attractive forces between its molecules in the liquid state are the very weak London dispersion forces. Both 2-propanol and 2-propanamine are polar molecules and associate in the liquid state by hydrogen bonding. Hydrogen bonding is stronger between alcohol molecules than between amine molecules because of the greater strength of an O-H----O hydrogen bond compared to an N-H----N hydrogen bond. It takes more energy (a higher temperature) to separate an alcohol molecule in the liquid state from its neighbors than to separate an amine molecule from its neighbors and, therefore, the alcohol has the higher boiling point.

16.19 Nitrogen is less electronegative than oxygen and, therefore, more willing to donate its unshared pair of electrons to H^+ in an acid-base reaction to form a salt.

16.21 (a) ethylammonium chloride   (b) diethylammonium chloride
(c) anilinium hydrogen sulfate

16.23 Structural formula A contains both an acid (the carboxyl group) and a base (the 1° amino group). The acid-base reaction between them gives structural formula B, which is the better representation of this amino acid.

16.25 Following are completed equations.

(a) 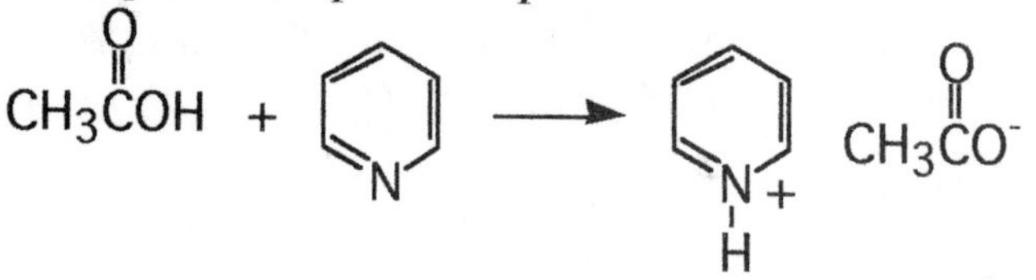 CH_3COH + [pyridine] → [N-H pyridinium] CH_3CO^-

(b)  [phenyl-CH2-CH-NH2] + HCl → [phenyl-CH2-CH-NH3+ Cl-]

(c) 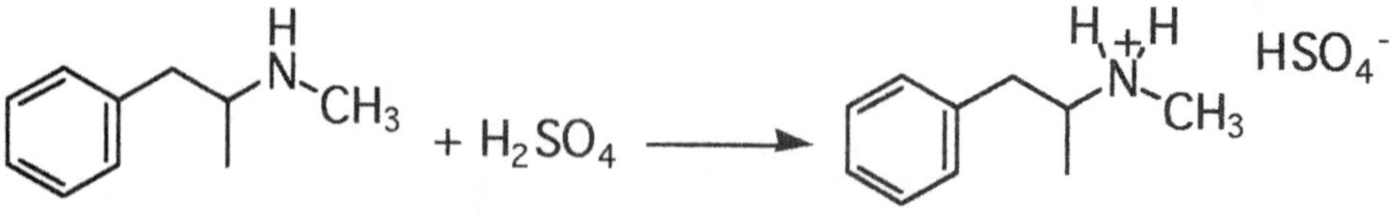 [phenyl-CH2-CH-NH-CH3] + H_2SO_4 → [phenyl-CH2-CH-N+(H)(H)-CH3 HSO_4^-]

16.27 The primary aliphatic amine is the stronger base and forms the salt with HCl. The salt is named pyridoxamine hydrochloride.

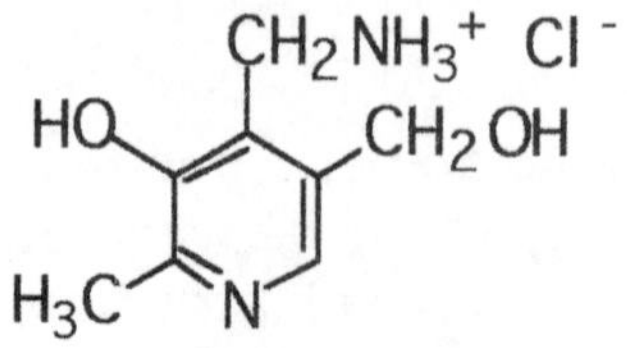

$CH_2NH_3^+$ Cl^-
HO, CH_2OH
H_3C N

16.29 Albuterol contains one 1° alcohol, one 2° alcohol, one phenol, and one 2° amine. Albuterol differs from epinephrine in that one phenolic -OH group of epinephrine is converted to a -CH$_2$OH group, and the N-methyl group of epinephrine is converted to an N-*tert*-butyl group.

16.31 Possible negative effects are long periods of sleeplessness, loss of weight, and paranoia.

16.33 Both coniine and nicotine have one stereocenter; two stereoisomers (one pair of enantiomers) are possible for each.

16.35 The four stereocenters of cocaine are marked with asterisks. Following is the structural formula of the salt formed by the reaction of cocaine with HCl.

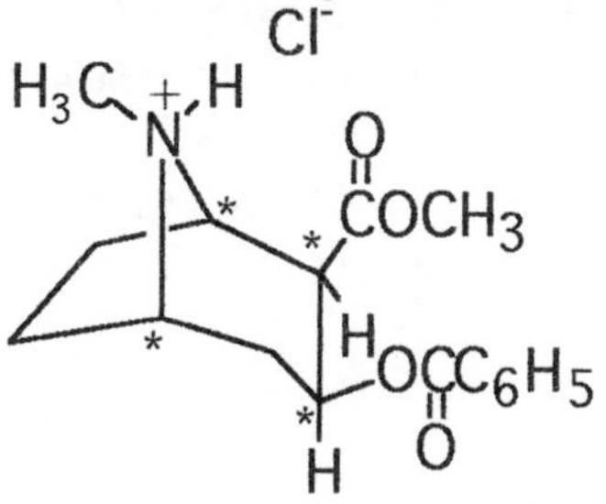

16.37 Neither Librium nor Valium is chiral.

16.39 No. No unreacted HCl is present.

16.41 Following is structural formula for each part.

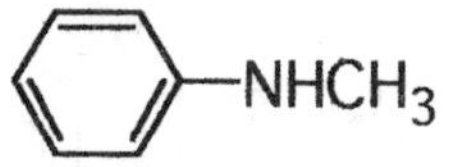

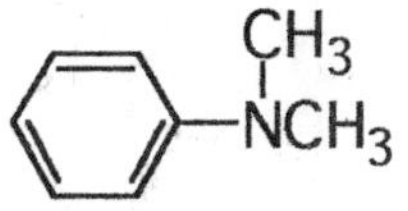

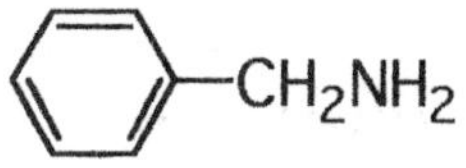

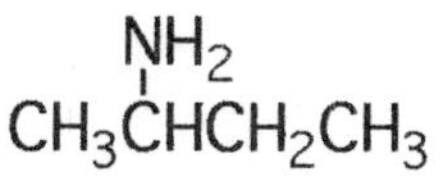

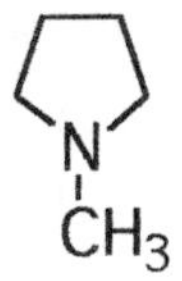

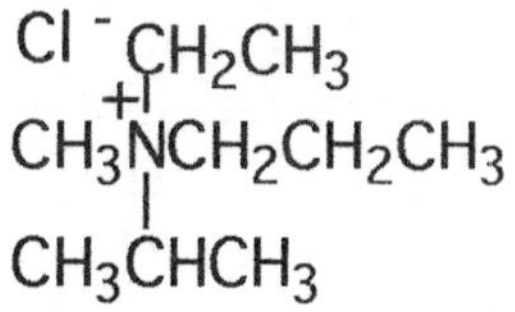

16.43 (a) CH$_3$SH is the strongest acid.   (b) (CH$_3$)$_2$NH is the strongest base.
(c) CH$_3$OH has the highest boiling point.
(d) Molecules of CH$_3$OH form the strongest hydrogen bonds.

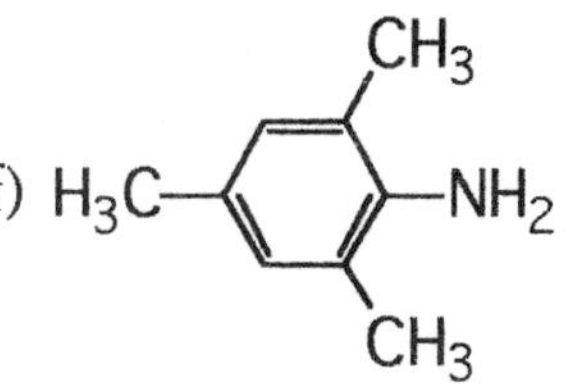

<u>16.45</u> Both alcohols and amines can interact with water by hydrogen bonding. Because amines and alcohols of the same molecular weight have about the same solubility in water, the strength of hydrogen bonding between their molecules must be comparable.

<u>16.47</u> (a) Following is its structural formula. (b) It has two stereocenters, marked with asterisks, and $2^2 = 4$ stereoisomers are possible.

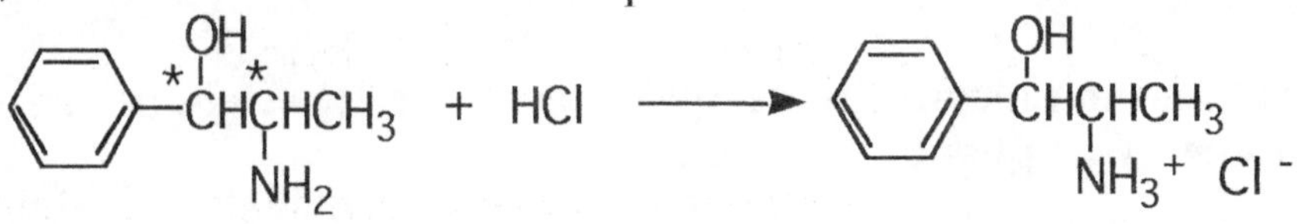

<u>16.49</u> Gabapentin is better represented by structural formula A, the internal salt. Structural formula B contains both an acid (-COOH) and a base ($-NH_2$) which will react by an internal proton-transfer reaction to form A.

<u>16.51</u> The 2° aliphatic amine is more basic than the heterocyclic aromatic amine. The three stereocenters are marked by asterisks.

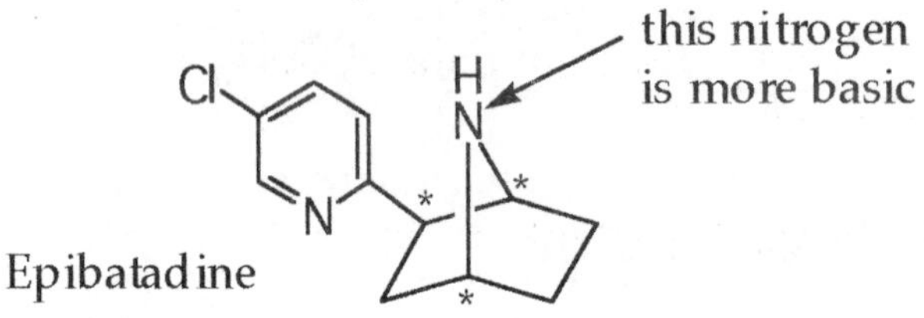

Chapter 17 Aldehydes and Ketones

17.1 (a) 3,3-dimethylbutanal   (b) cyclopentanone   (c) (S)-2-phenylpropanal

17.2 Following are line-angle formulas for each aldehyde with the molecular formula $C_6H_{12}O$. In the three that are chiral, the stereocenter is marked by an asterisk.

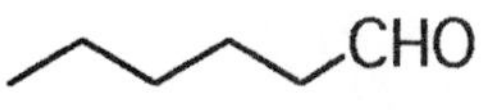

Hexanal 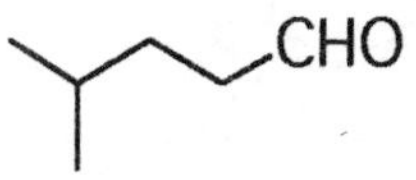   4-Methylpentanal 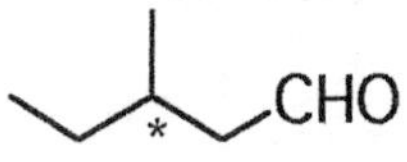   3-Methylpentanal 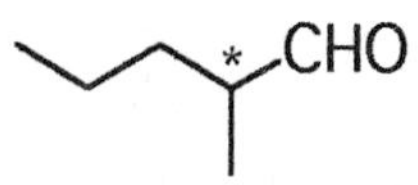   2-Methypentanal

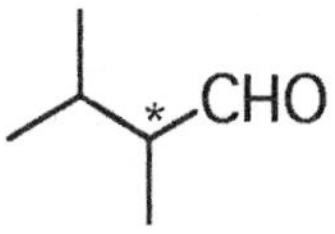

2,3-Dimethylbutanal 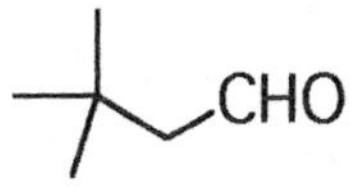   3,3-Dimethylbutanal    2,2-Dimethylbutanal

17.3 (a) 2,3-dihydroxypropanal   (b) 2-aminobenzaldehyde
(c) 5-amino-2-pentanone

17.4 Each aldehyde is oxidized to a carboxylic acid.

(a) 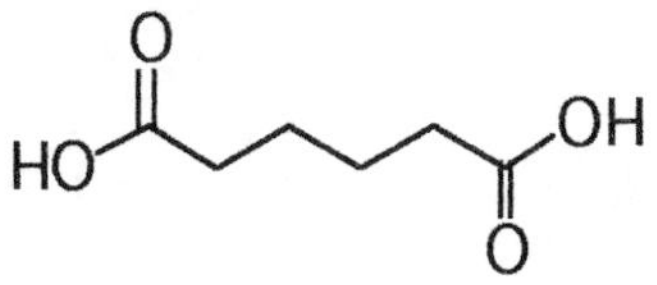   (b)

Hexanedioic acid   3-Phenylpropanoic acid
(Adipic acid)

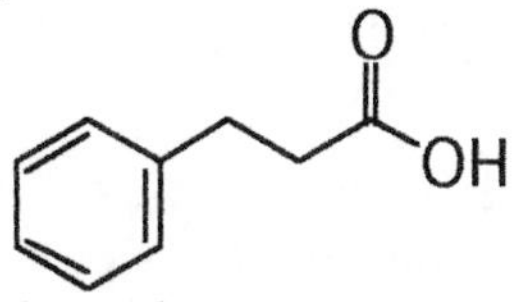

17.5 Each primary alcohol comes from reduction of an aldehyde. Each secondary alcohol comes from reduction of a ketone.

(a) 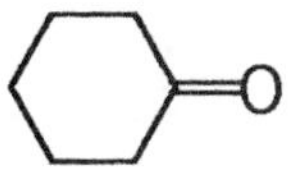   (b) 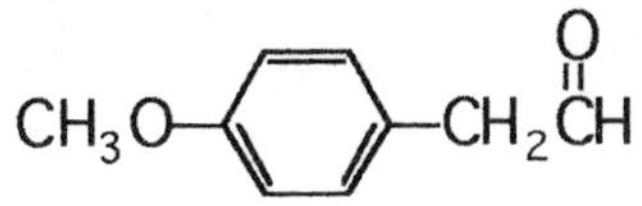 CH_3O-⬡$-CH_2\overset{O}{\overset{\|}{C}}H$ (c)

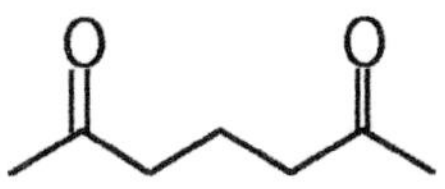

17.6 Shown first is the hemiacetal and then the acetal.

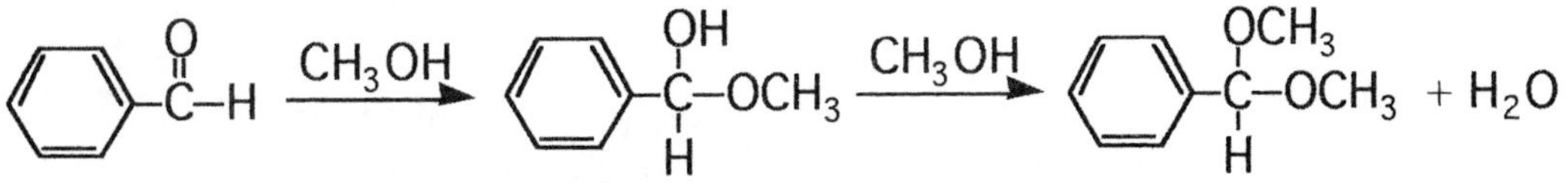

Benzaldehyde   Hemiacetal   Acetal

17.7 (a) A hemiacetal formed from 3-pentanone (a ketone) and ethanol.

93

(b) Neither a hemiacetal nor an acetal. This compound is the dimethyl ether of
ethylene glycol.
(c) An acetal derived from 5-hydroxypentanal and methanol.

17.8 Following is the keto form of each enol.

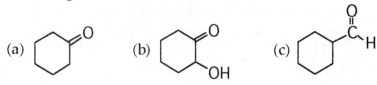

(a)　　　　(b)　　　　(c)

17.9 Oxidation of a primary alcohol gives either an aldehyde or a carboxylic acid,
depending on the experimental conditions. Oxidation of a secondary alcohol gives
a ketone.

(a) 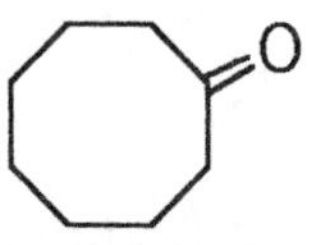　　　　(b) 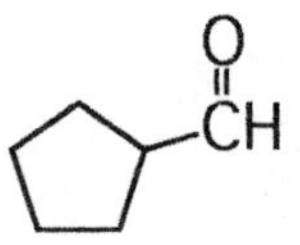　　　　or

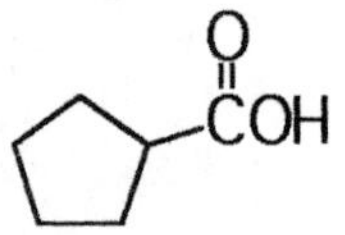

17.11 The carbonyl carbon of an aldehyde is bonded to at least one hydrogen. The
carbonyl carbon of a ketone is bonded to two carbon groups.

17.13 No. To be a carbon stereocenter, the carbon atom must have four different groups
bonded to it. The carbon atom of a carbonyl group has only three groups bonded
to it.

17.15 (a) Cortisone contains three ketones, one 3° alcohol, one 1° alcohol, and one
carbon-carbon double bond.
(b) Aldosterone contains two ketones, one aldehyde, one 1° alcohol, one 2°
alcohol, and one carbon-carbon double bond.

17.17 Following are structural formulas for each aldehyde.

(a) 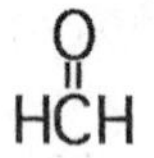 HCH　(b) 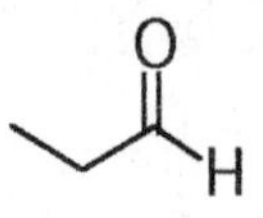　(c)

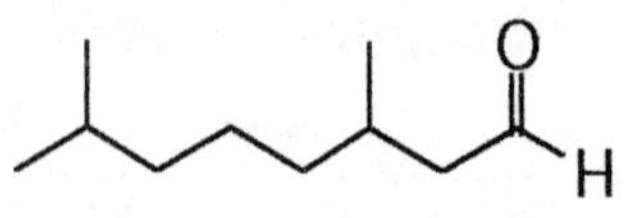

(d) 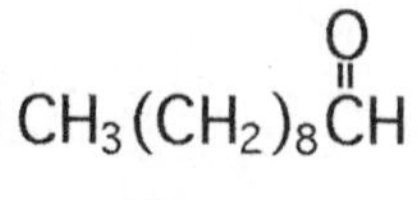 $CH_3(CH_2)_8CH$　(e) 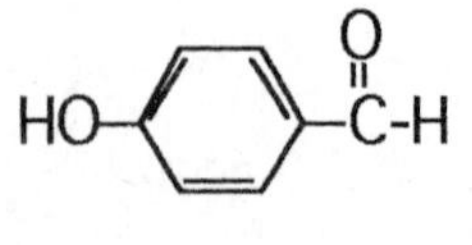 HO-　-C-H　(f)

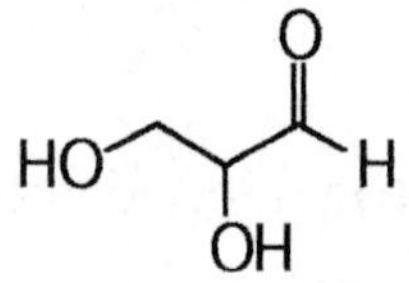

17.19 (a) 4-heptanone　　　(b) 2-methylcyclopentanone
(c) cis-2-methyl-2-butenal　(d) (S)-2-hydroxypropanal
(e) 1-phenyl-2-propanone　(f) hexanedial

17.21 (a) The chain is not numbered correctly. Its name is 2-butanone.

(b) The compound is an aldehyde. Its name is butanal.
(c) The longest chain is five carbons. Its name is pentanal.
(d) The location of the ketone takes precedence. Its name is 3,3-dimethyl-2-butanone.

17.23 (a) ethanol   (b) 3-pentanone   (c) butanal   (d) 2-butanol

17.25 Acetone has the higher boiling point because of the intermolecular attraction between the carbonyl groups of acetone molecules.

17.27 Acetaldehyde forms hydrogen bonds with water primarily through its carbonyl oxygen.

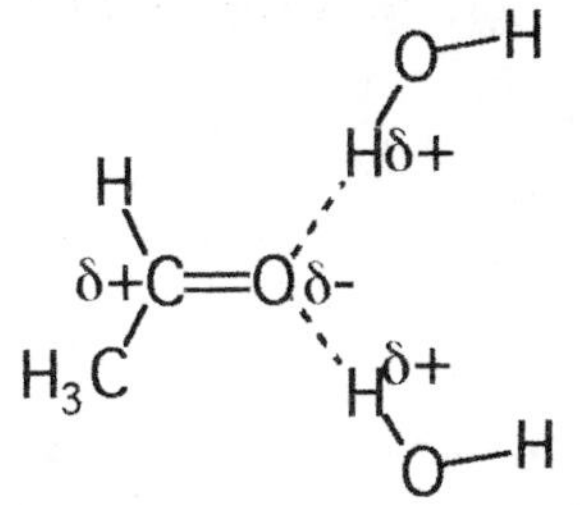

17.29 Aldehydes are oxidized by this reagent to carboxylic acids. Ketones are not oxidized under these conditions. Secondary alcohols are oxidized to ketones.

(a) $CH_3CH_2CH_2COH$ with =O   (b) 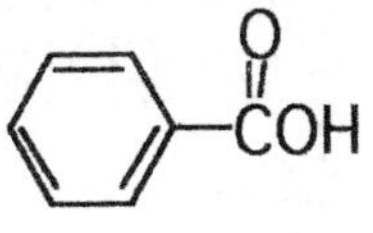 benzene ring–COH with =O   (c) no reaction   (d) 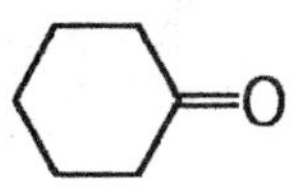 cyclohexanone

17.31 (a) Treat each with Tollens' reagent. Only pentanal will give a silver mirror.
      (b) Treat each with $K_2Cr_2O_7/H_2SO_4$. Only 2-pentanol is oxidized (to 2-pentanone), which causes the red color of $Cr_2O_7^{2-}$ ion to disappear and be replaced by the green color of Cr^{3+} ion.

17.33 The white solid is benzoic acid, formed by air oxidation of benzaldehyde.

17.35 These experimental conditions reduce an aldehyde to a primary alcohol and a ketone to a secondary alcohol.

(a) $CH_3CHCH_2CH_3$ with OH   (b) $CH_3(CH_2)_4CH_2OH$ (c) 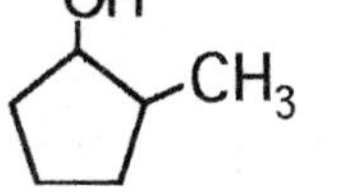 2-methylcyclopentanol   (d) 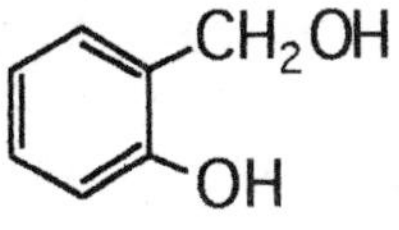 benzene ring with CH_2OH and OH

17.37 (a) Following is its structural formula.

$$\underset{\substack{\text{1,3-Dihydroxy-2-propanone}\\\text{(Dihydroxyacetone)}}}{HOCH_2\overset{\overset{O}{\|}}{C}CH_2OH} + H_2 \xrightarrow[\text{catalyst}]{\text{metal}} \underset{\substack{\text{1,2,3-Propanetriol}\\\text{(Glycerol, glycerin)}}}{HOCH_2\overset{\overset{OH}{|}}{C}HCH_2OH}$$

(b) Because it has two hydroxyl groups and one carbonyl group, all of which can interact with water molecules by hydrogen bonding, predict that it is soluble in water. (c) Its reduction gives 1,2,3-propanetriol, better known as glycerol or glycerin.

17.39 The first two reactions are reduction. There is no reaction in (c) or (d); ketones are not oxidized by these reagents.

(a,b) [structure: benzene ring with $-\overset{\overset{OH}{|}}{C}HCH_3$]

17.41 (a) an acetal   (b) a hemiacetal   (c) an acetal
     (d) an acetal   (e) an acetal   (f) neither

17.43 Following are equations for the formation of each hemiacetal and acetal.

(a) $CH_3CH_2\overset{\overset{O}{\|}}{C}\text{-H} \xrightarrow{CH_3OH} CH_3CH_2\overset{\overset{OH}{|}}{\underset{\underset{OCH_3}{|}}{C}}\text{-H} \xrightarrow{CH_3OH} CH_3CH_2\overset{\overset{OCH_3}{|}}{\underset{\underset{OCH_3}{|}}{C}}\text{-H} + H_2O$

(b) [cyclopentanone] $=O \xrightarrow{CH_3OH}$ [cyclopentane with OH and OCH₃] $\xrightarrow{CH_3OH}$ [cyclopentane with OCH₃ and OCH₃] $+ H_2O$

17.45 *Hydration* refers to the addition of one or more molecules of water to a substance. An example of hydration is the acid-catalyzed addition of water to propene to give 2-propanol. *Hydrolysis* refers to the reaction of a substance with water, generally with the breaking (lysis) of one or more bonds in the substance. An example of hydrolysis is the acid-catalyzed reaction of an acetal with a molecule of water to give an aldehyde or ketone and two molecules of alcohol.

17.47 Compounds (a), (b), (d), and (f) undergo keto-enol tautomerism.

17.49 Following are the keto forms of each enol.

(a) [cyclopentanone] $=O$ (b) $CH_3\overset{\overset{O}{\|}}{C}CH_2CH_2CH_2CH_3$ (c) [benzene]$-CH_2\overset{\overset{O}{\|}}{C}CH_3$

17.51 Compounds (a), (b), and (c) can be formed by reduction of the aldehyde or ketone shown. Compound (d) is a 3° alcohol and cannot be formed in this manner.

(a) $CH_3\overset{O}{\overset{\|}{C}}CH_3$ (b) 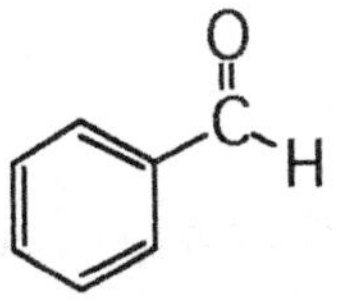 [benzaldehyde structure] $\overset{O}{\overset{\|}{C}}{}^{\backslash}H$ (c) $H-\overset{O}{\overset{\|}{C}}-H$

17.53 The alkene of three carbons is propene. Acid-catalyzed hydration of this alkene gives 2-propanol as the major product, not 1-propanol.

17.55 Each conversion can be brought about by acid-catalyzed hydration of the alkene to a secondary alcohol, followed by oxidation of the secondary alcohol to a ketone.

(a) $CH_2=CHCH_2CH_2CH_3 \xrightarrow[H_2SO_4]{H_2O} CH_3\overset{OH}{\underset{|}{C}}HCH_2CH_2CH_3 \xrightarrow[H_2SO_4]{K_2Cr_2O_7} CH_3\overset{O}{\overset{\|}{C}}CH_2CH_2CH_3$

(b) 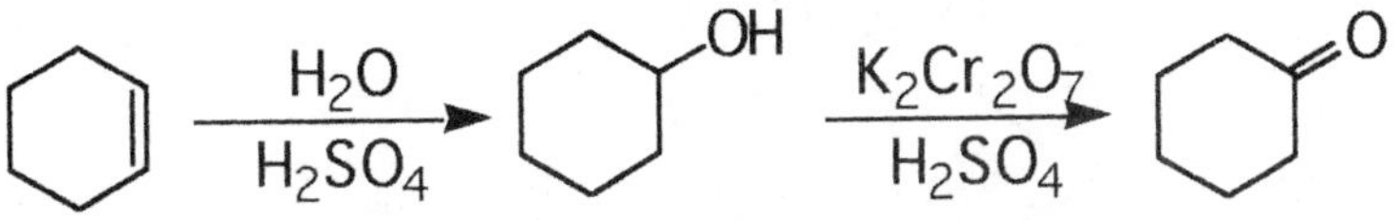 [cyclohexene structure] $\xrightarrow[H_2SO_4]{H_2O}$ [cyclohexanol structure] $\xrightarrow[H_2SO_4]{K_2Cr_2O_7}$ [cyclohexanone structure]

17.57 The aldehyde and ketone functional groups are circled.

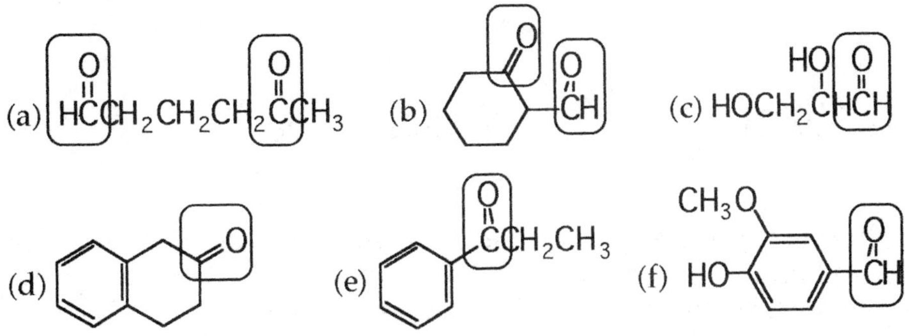

(a) $H\overset{O}{\overset{\|}{C}}CH_2CH_2CH_2\overset{O}{\overset{\|}{C}}CH_3$ (b) [cyclohexanone with CH group structure]   (c) $HOCH_2\overset{HO}{\underset{|}{C}}H\overset{O}{\overset{\|}{C}}H$

(d) [bicyclic ketone structure]   (e) [phenyl ketone $\overset{O}{\overset{\|}{C}}CH_2CH_3$ structure]   (f) $HO-$[aromatic ring with CH_3O]$-\overset{O}{\overset{\|}{C}}H$

17.59 Each compound reduces a carbonyl group of an aldehyde or ketone to an alcohol by delivering a hydride ion (H:⁻) to the carbonyl carbon.

17.61 Formulas for the one ketone and two aldehydes with molecular formula C_4H_8O are

(a) $CH_3\overset{O}{\overset{\|}{C}}CH_2CH_3$ (b) $CH_3CH_2CH_2\overset{O}{\overset{\|}{C}}H$ and $CH_3\overset{O}{\underset{\underset{CH_3}{|}}{C}}H\overset{\|}{C}H$

17.63 2-Propanol has the higher boiling point because of the greater attraction between its molecules due to hydrogen bonding through its hydroxyl group.

17.65 (a) Treat each compound with Tollens' reagent. Only benzaldehyde reduces Ag^+ to give a precipitate of silver metal as a silver mirror.
(b) Treat each compound with Tollens' reagent as in part (a). Only acetaldehyde reduces Ag^+ to give a precipitate of silver metal as a silver mirror.

17.67 Shown in the equation are structural formulas for the equilibrium products.

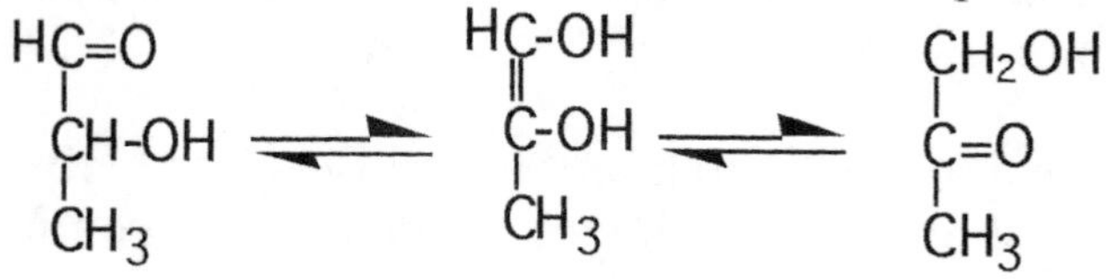

$$
\begin{array}{ccc}
\overset{HC=O}{\underset{\underset{CH_3}{|}}{\overset{|}{CH\text{-}OH}}} & \rightleftharpoons \quad \overset{HC\text{-}OH}{\underset{\underset{CH_3}{|}}{\overset{\|}{C\text{-}OH}}} \rightleftharpoons & \overset{CH_2OH}{\underset{\underset{CH_3}{|}}{\overset{|}{C=O}}}
\end{array}
$$

An α-hydroxyaldehyde   An enediol   An α-hydroxyketone

17.69 (a) Carbon 5 provides the -OH group, and carbon 1 provides the -CHO group. (b) Following is a structural formula for the free aldehyde.

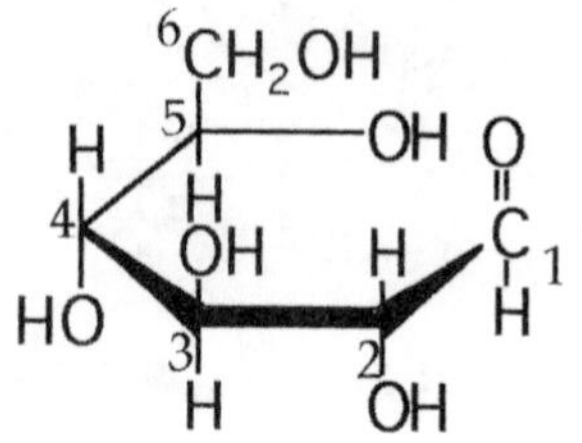

Chapter 18 Carboxylic Acids and Their Derivatives

<u>18.1</u> (a) 2,3-dihydroxypropanoic acid   (b) 3-aminopropanoic acid
(c) 3,5-dihydroxy-3-methylpentanoic acid

<u>18.2</u> Each acid is converted to its ammonium salt.

(a) $CH_3(CH_2)_2COOH + NH_3 \longrightarrow CH_3(CH_2)_2COO^- NH_4^+$

   Butanoic acid   Ammonium butanoate
   (Butyric acid)   (Ammonium butyrate)

(b) $CH_3\overset{OH}{\underset{|}{C}}HCOOH + NH_3 \longrightarrow CH_3\overset{OH}{\underset{|}{C}}HCOO^- NH_4^+$

   2-Hydroxypropanoic   Ammonium 2-hydroxypropanoate
        acid   (Ammonium lactate)
   (Lactic acid)

<u>18.3</u> (a) ethyl benzoate   (b) phenyl acetate

<u>18.4</u> Following is a structural formula for each amide.

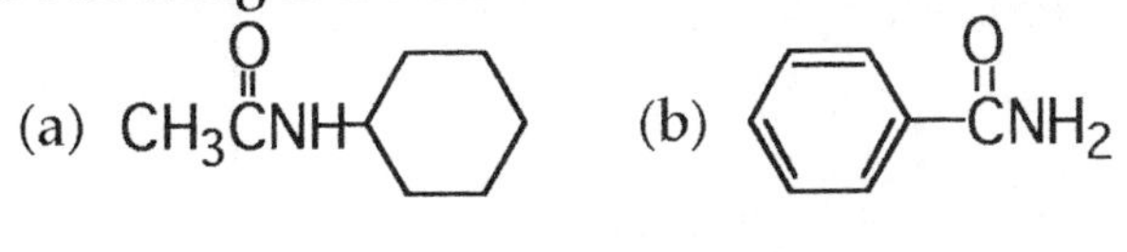

(a) $CH_3\overset{O}{\overset{||}{C}}NH-$⬡   (b) ⬡$-\overset{O}{\overset{||}{C}}NH_2$

<u>18.5</u> Following is the structural formula of each ester.

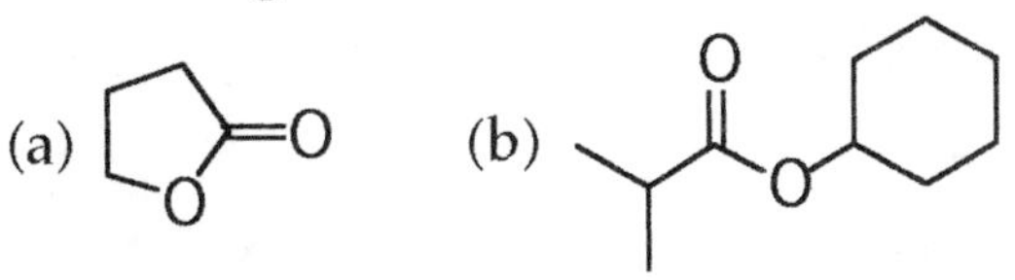

(a) [lactone structure]=O   (b) [structure]

<u>18.6</u> Under basic conditions, as in part (a), each carboxyl group is present as a carboxylic anion. Under acidic conditions, as in part (b), each carboxyl group is present in its un-ionized form.

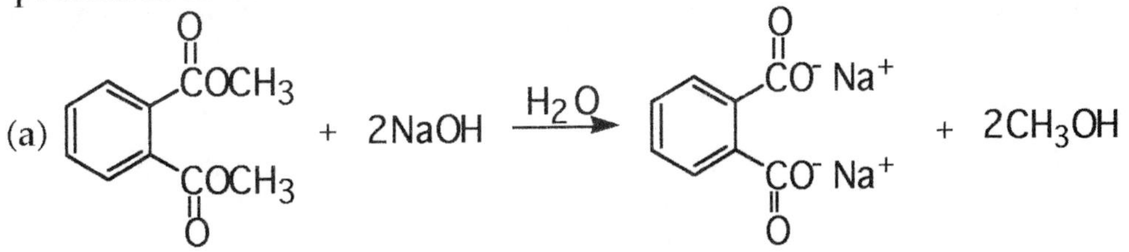

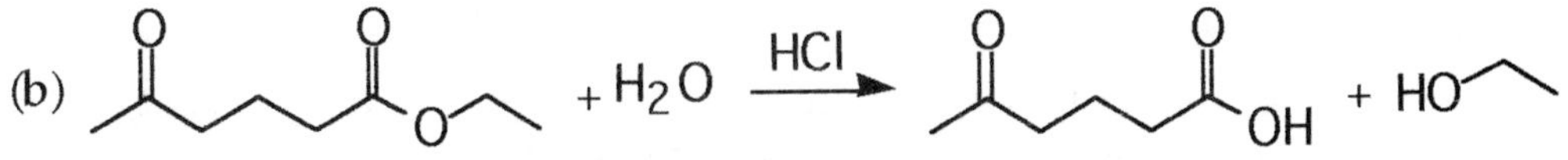

(a) [aromatic diester with two $COCH_3$ groups] + 2NaOH $\xrightarrow{H_2O}$ [aromatic salt with two $CO^- Na^+$ groups] + $2CH_3OH$

(b) [diketone ester structure] + H_2O $\xrightarrow{HCl}$ [diketone acid structure] $_{OH}$ + HO⌐

<u>18.7</u> In aqueous NaOH, each carboxyl group is present as a carboxylic anion, and each amine is present in its unprotonated form.

(a) $CH_3\overset{O}{\overset{||}{C}}N(CH_3)_2 + NaOH \xrightarrow[\text{heat}]{H_2O} CH_3\overset{O}{\overset{||}{C}}O^- Na^+ + (CH_3)_2NH$

(b) [structure: 2-piperidinone ring with C=O and NH] + NaOH $\xrightarrow[\text{heat}]{\text{H}_2\text{O}}$ H$_2$N‒[chain]‒C(=O)O$^-$Na$^+$

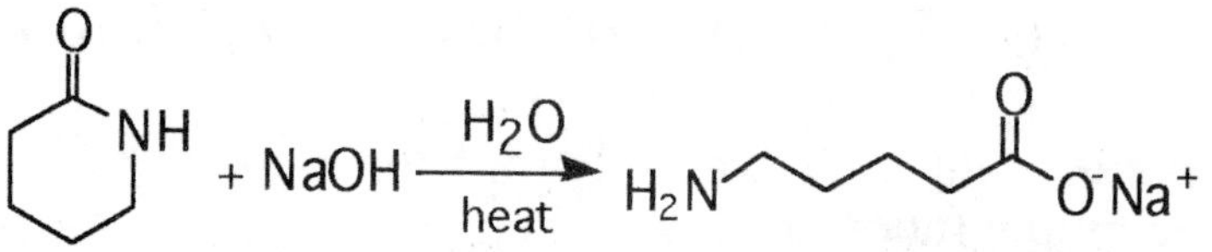

18.8 Of these four carboxylic acids with molecular formula $C_5H_{10}O_2$, only one is chiral. Its stereocenter is marked by an asterisk.

[structures] COOH COOH *COOH COOH

Pentanoic acid   3-Methylbutanoic   2-Methylbutanoic   2,2-Dimethyl-
acid   acid   propanoic acid

18.9 (a) 3,4-dimethylpentanoic acid   (b) 2-aminobutanoic acid
(c) hexanoic acid

18.11 Following are structural formulas for each carboxylic acid.

(a) [structure: para-nitrobenzyl COOH, O$_2$N on ring]   (b) H$_2$N‒[chain]‒COOH

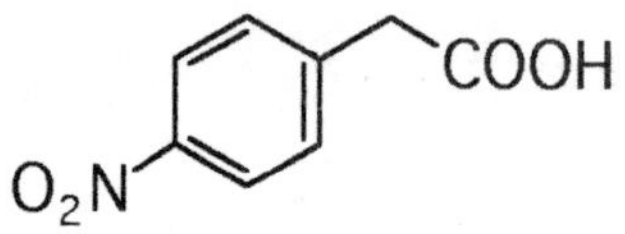

(c) [structure: phenyl propyl COOH]   (d) [structure: cyclohexene with two COOH groups]

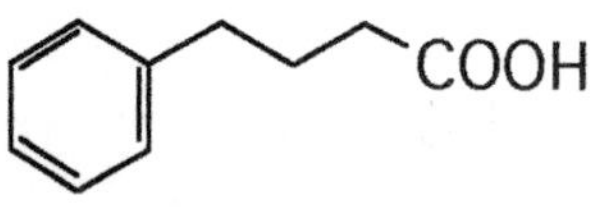

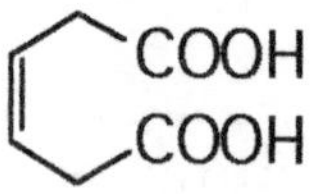

18.13 Following are structural formulas for each salt.

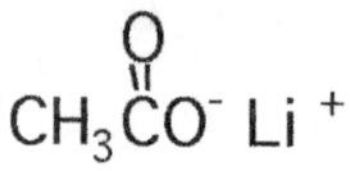

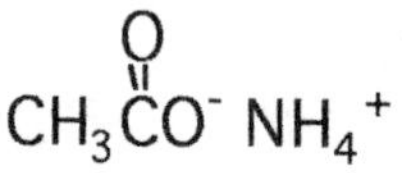

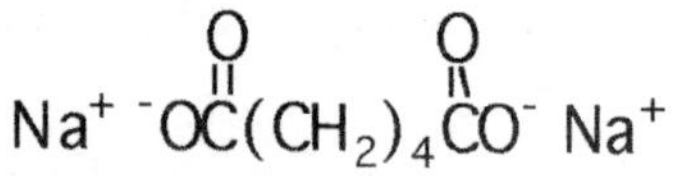

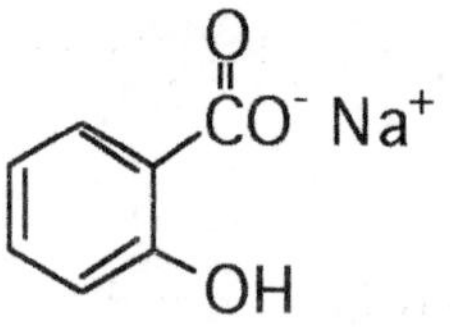

(a) [structure: phenyl‒CO$^-$ Na$^+$]   (b) CH$_3$CO$^-$ Li$^+$ (c) CH$_3$CO$^-$ NH$_4$$^+$

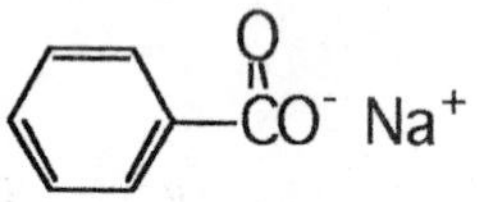

(d) Na$^+$ $^-$OC(CH$_2$)$_4$CO$^-$ Na$^+$ (e) [structure: benzene ring with CO$^-$ Na$^+$ and OH]   (f) (CH$_3$CH$_2$CH$_2$CO$^-$)$_2$Ca^{2+}

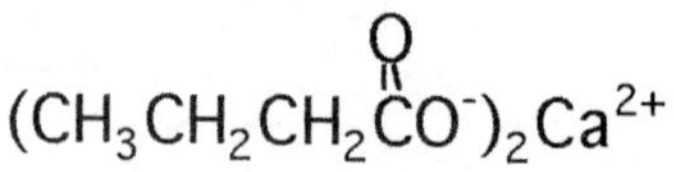

18.15 One of the carboxyl groups in this salt is present as -COO$^-$, the other as -COOH.

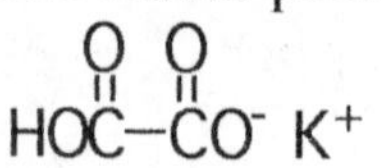

HOC‒CO$^-$ K$^+$

18.17 If you draw this molecule correctly to show this internal hydrogen bonding, you will see that the hydrogen-bonded part of the molecule forms a six-membered ring.

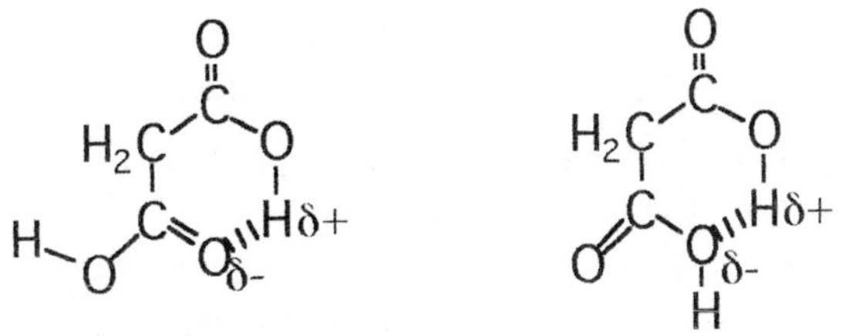

18.19 In order of increasing boiling point, they are heptanal (bp 153°C), 1-heptanol (bp 176°C), and heptanoic acid (bp 223°C).

18.21 In order of increasing solubility in water, they are decanoic acid, pentanoic acid, and acetic acid.

18.23 Following are structural formulas for the indicated starting material.

(a) $CH_3(CH_2)_4CH_2OH$ (b) $CH_3(CH_2)_4\overset{\overset{O}{\|}}{C}H$ (c) $HOCH_2(CH_2)_4CH_2OH$
      $C_6H_{14}O$ $C_6H_{12}O$ $C_6H_{14}O_2$

18.25 In order of increasing acidity, they are benzyl alcohol, phenol, and benzoic acid.

18.27 Following are completed equations for these acid-base reactions.

(a) 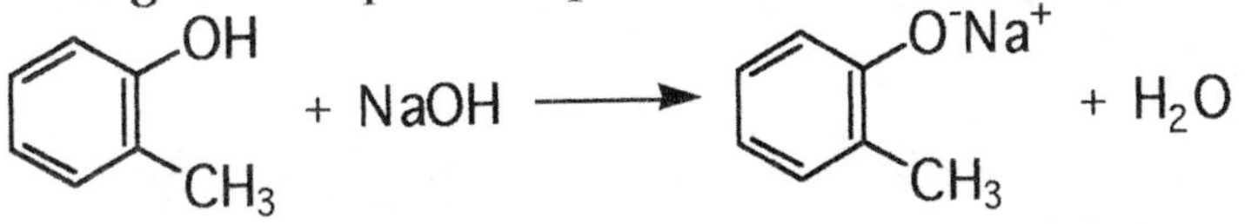 [structure: 2-methylphenol] + NaOH $\longrightarrow$ [structure: sodium 2-methylphenolate] + H_2O

(b) 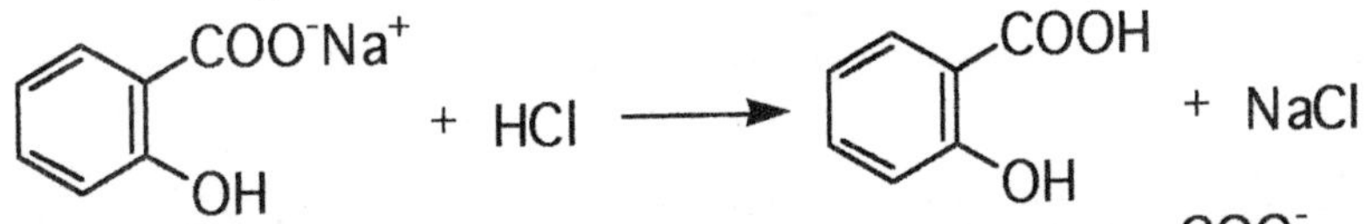 [structure: sodium salicylate] + HCl $\longrightarrow$ [structure: salicylic acid] + NaCl

(c) 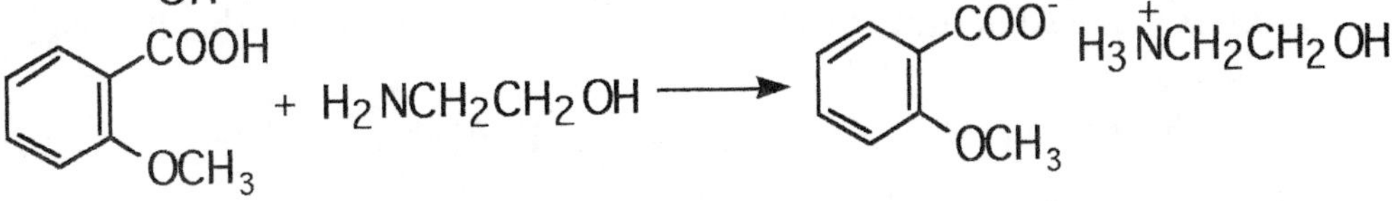 [structure: 2-methoxybenzoic acid] + $H_2NCH_2CH_2OH$ $\longrightarrow$ [structure: ammonium carboxylate salt] $H_3\overset{+}{N}CH_2CH_2OH$

(d) 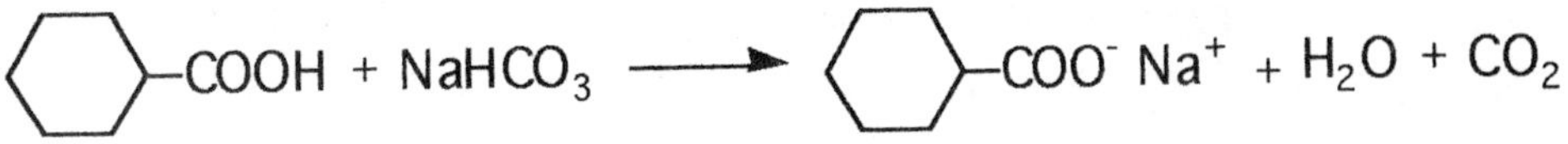 [cyclohexyl]$-COOH$ + $NaHCO_3$ $\longrightarrow$ [cyclohexyl]$-COO^-\ Na^+$ + H_2O + CO_2

18.29 CH_3COOH at pH 2.0, equal amounts of CH_3COOH and CH_3COO^- at pH 4.76, and CH_3COO^- at pH 8.0 or higher.

18.31 The pK_a of ascorbic acid is 4.10. At this pH, ascorbic acid is present 50% as ascorbic acid and 50% as ascorbate ion. At pH 7.35 to 7.45, which is more basic than pH 4.10, ascorbic acid would be present as ascorbate ion.

18.33 In part (a), the -COOH group is a stronger acid than the $-NH_3^+$ group.
(a) $CH_3\underset{\overset{|}{NH_3^+}}{CH}COOH$ + NaOH $\longrightarrow$ $CH_3\underset{\overset{|}{NH_3^+}}{CH}COO^-Na^+$ + H_2O

101

(b) $CH_3\underset{\underset{NH_3^+}{|}}{C}HCOO^- Na^+$ + NaOH $\longrightarrow$ $CH_3\underset{\underset{NH_2}{|}}{C}HCOO^- Na^+$ + H_2O

18.35 In part (a), the amine is the stronger base.

(a) $CH_3\underset{\underset{NH_2}{|}}{C}HCOO^- Na^+$ + HCl $\longrightarrow$ $CH_3\underset{\underset{NH_3^+}{|}}{C}HCOO^- Na^+$ + Cl$^-$

(b) $CH_3\underset{\underset{NH_3^+}{|}}{C}HCOO^- Na^+$ + HCl $\longrightarrow$ $CH_3\underset{\underset{NH_3^+}{|}}{C}HCOOH$ + NaCl

18.37 Following is a structural formula for the ester formed in each reaction.

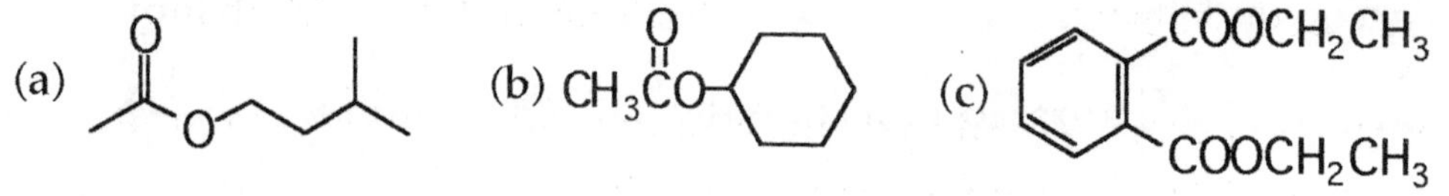

(a) (b) $CH_3\overset{O}{\overset{||}{C}}O$— (c)

18.39 Following are structural formulas for acid and alcohol from which each ester is derived.

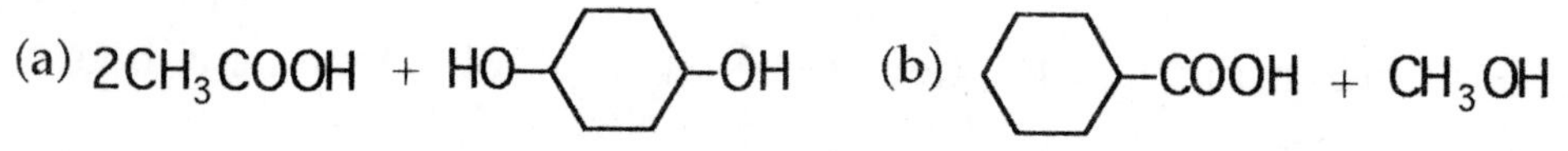

(a) $2CH_3COOH$ + HO——OH (b) —COOH + CH_3OH

(c) $2CH_3OH$ + $HOOCCH_2CH_2COOH$ (d) 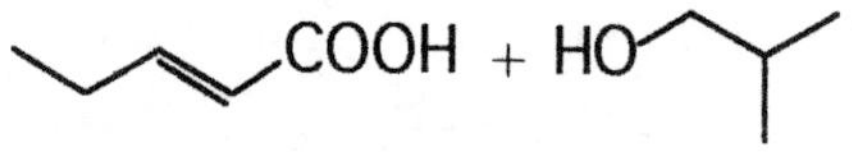 COOH + HO

18.41 Following is a structural formula for methyl 4-hydroxybenzoate.

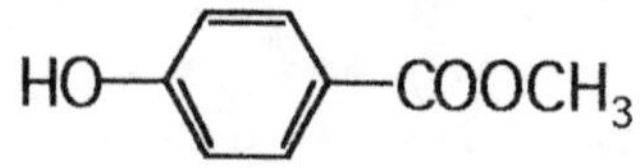

HO——COOCH$_3$

18.43 Following are structural formulas for the reagents to synthesize each amide.

(a) 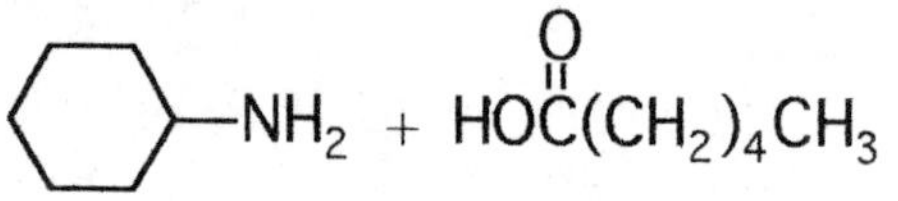 —NH$_2$ + $HO\overset{O}{\overset{||}{C}}(CH_2)_4CH_3$ (b) $(CH_3)_2CH\overset{O}{\overset{||}{C}}OH$ + $HN(CH_3)_2$

(c) $2NH_3$ + $HO\overset{O}{\overset{||}{C}}(CH_2)_4\overset{O}{\overset{||}{C}}OH$

18.45 (a) Both lidocaine and mepivacaine contain an amide and a 3° amine. (b) Both are derived from 2,6-dimethylaniline. In addition, the aliphatic amine nitrogen in each is separated by one carbon from the carbonyl group of the amide.

18.47 Following is a structural formula for each synthetic flavoring agent.

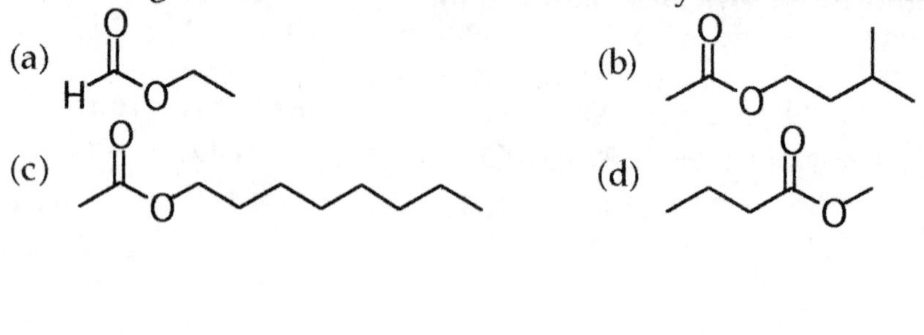

(a) (b)

(c) (d)

(e)

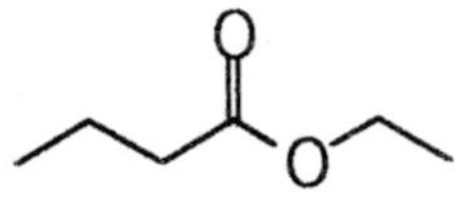

(f)

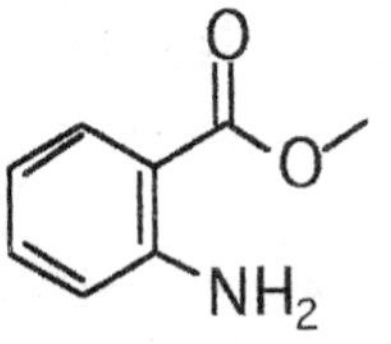

18.49 Saponification is the hydrolysis of an ester using aqueous NaOH or KOH to give the sodium or potassium salt of the carboxylic acid and an alcohol. Following is a balanced equation for the saponification of methyl acetate.

$$CH_3\overset{O}{\overset{\|}{C}}OCH_3 + NaOH \xrightarrow{H_2O} CH_3\overset{O}{\overset{\|}{C}}O^- Na^+ + CH_3OH$$

Methyl acetate   Sodium acetate   Methanol

18.51 Each reaction brings about hydrolysis of the amide bond. Each product is shown as it would exist under the specified reaction conditions.

(a)

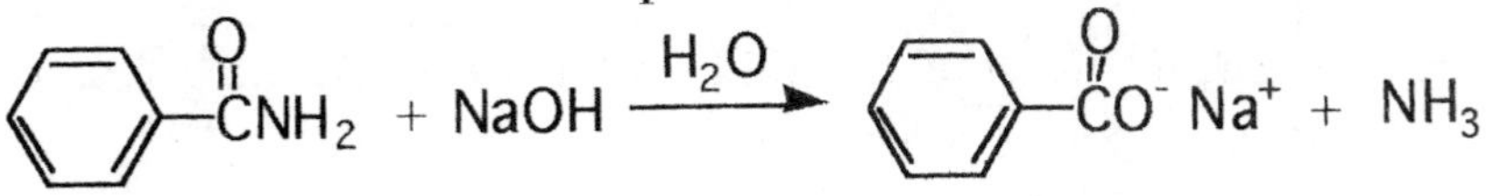

(b)

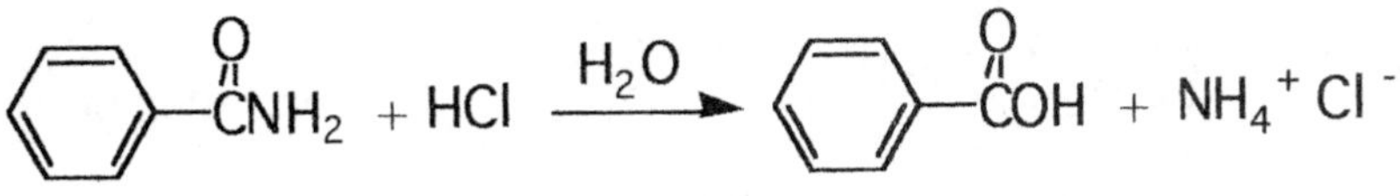

18.53 The products in each part are an amide and acetic acid.

(a)

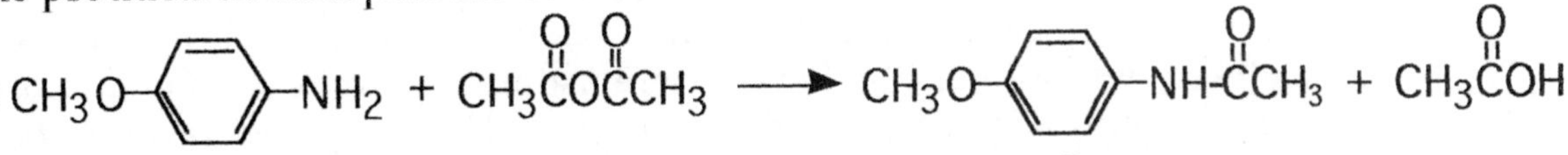

(b)

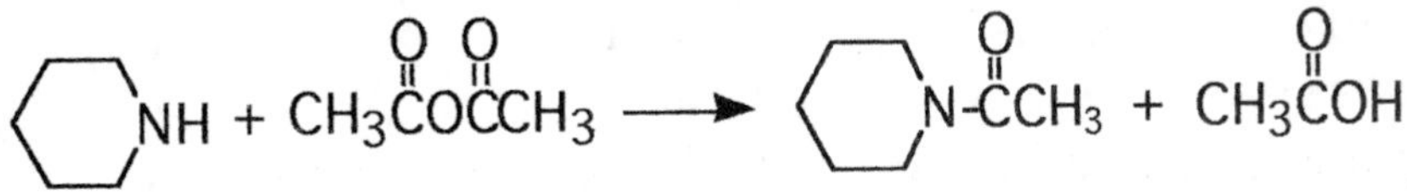

18.55 (a) Phenobarbital contains four amide groups. (b) Complete hydrolysis of all amide bonds gives a dicarboxylic acid dianion, two moles of ammonia, and one mole of sodium carbonate.

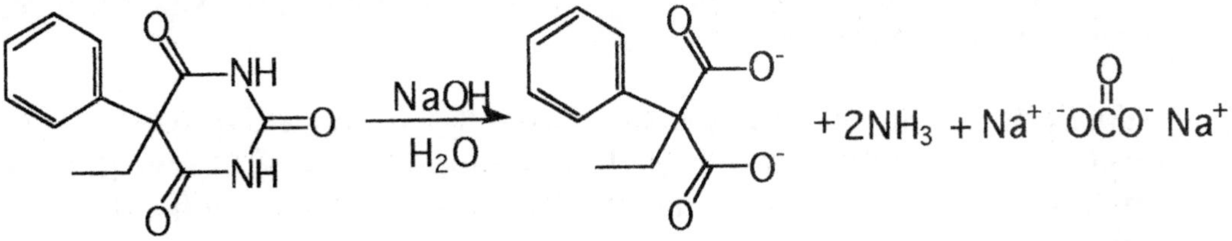

18.57 In nylon-66 and Kevlar, the monomer units are joined by amide bonds.

18.59 In Dacron and Mylar, the monomer units are joined by ester bonds.

18.61 Following are structural formulas for the mono-, di-, and triethyl esters.

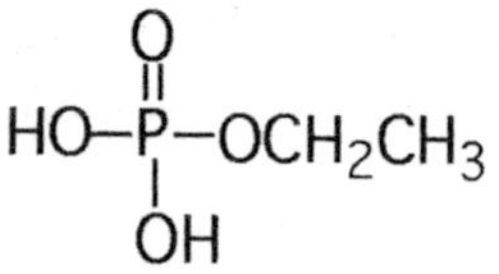

Ethyl phosphate

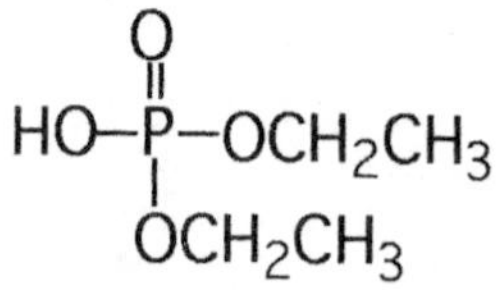

Diethyl phosphate

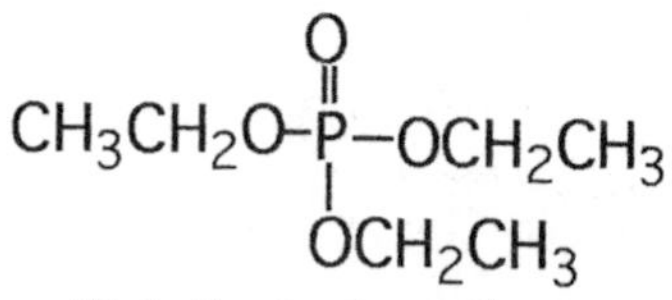

Triethyl phosphate

18.63 Two molecules of water are split out.

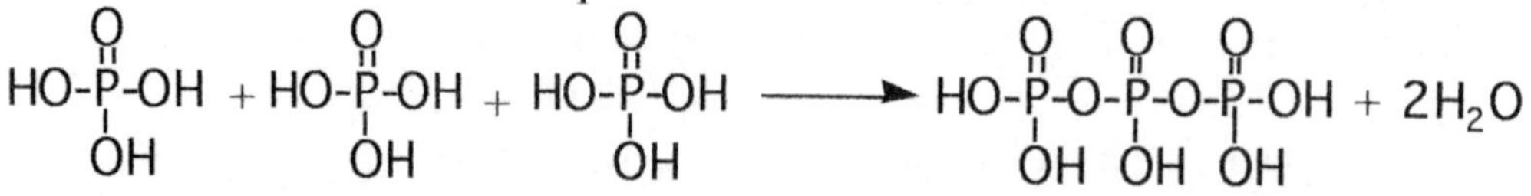

18.65 The arrow points to the ester group. On the right is chrysanthemic acid.

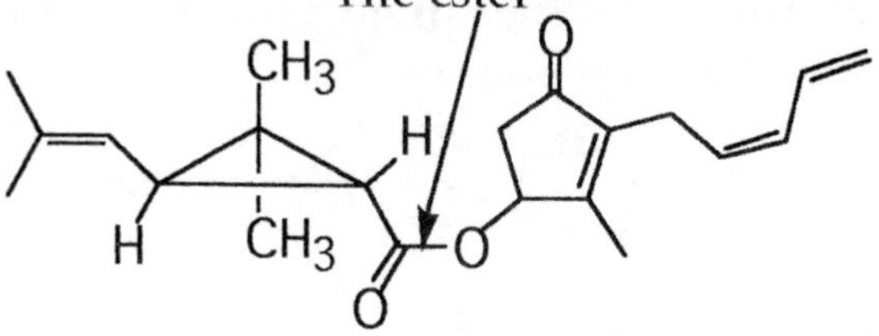

The ester

Pyrethrin I

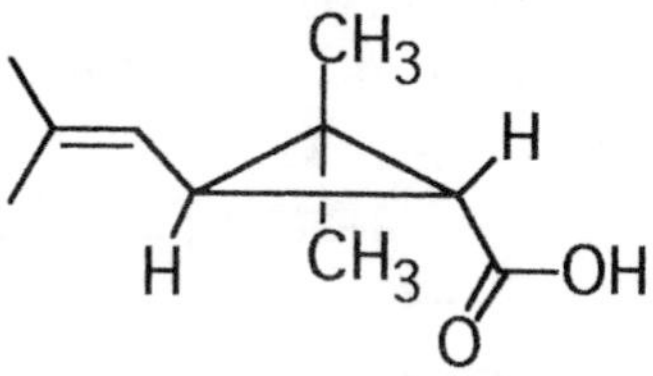

Chrysanthemic acid

18.67 (a) The *cis/trans ratio* refers to the cis-trans relationship between the ester group and the carbon-carbon double bond in the three-membered ring. (b) Permethrin has three stereocenters, and eight stereoisomers (four pairs of enantiomers) are possible for it. The designation "(+/-)" refers to the fact that the members of each pair of possible enantiomers are present in equal amounts; that is, each pair of enantiomers is present as a racemic mixture.

18.69 The compound is salicin. Removal of the glucose unit and oxidation of the primary alcohol to a carboxylic acid gives salicylic acid.

18.71 The moisture present in humid air may be sufficient to bring about hydrolysis of the ester to yield salicylic acid and acetic acid. The vinegar-like odor is due to the presence of acetic acid.

18.73 A *sunblock* prevents all ultraviolet radiation from reaching protected skin by reflecting it away from the skin. A *sunscreen* prevents a portion of the ultraviolet radiation from reaching protected skin. Its effectiveness is related to its skin protection factor (SPF).

18.75 They all contain an ester bonded to an alkyl group as well as a benzene ring. The benzene has either a nitrogen atom or an oxygen atom on it.

18.77 Lactomer stitches dissolve as the ester groups in the polymer chain are hydrolyzed until only glycolic acid and lactic acid remain. These small molecules are metabolized and excreted by existing biochemical pathways.

18.79 Propanoic acid has a boiling point of 141°C, and methyl acetate a boiling point of 57°C. The boiling point of propanoic acid is higher because of the association of adjacent molecules by hydrogen bonding.

18.81 In a solution of pH 4.07, lactic acid is 50 percent in the un-ionized (lactic acid) form and 50 percent in the ionized (lactate anion) form. In gastric juice of pH 1.0 to 3.0, which is more acidic than pH 4.10, lactic acid is present primarily as un-ionized lactic acid molecules.

18.83 Following is an equation for this synthesis of acetaminophen.

Acetic anhydride   4-Aminophenol   Acetaminophen

18.85 Hydrolysis gives one mole of 2,3-dihydroxypropanoic acid and two moles of phosphoric acid. At pH 7.35 - 7.45, the carboxyl group is present as its anion, and phosphoric acid is present as its dianion.

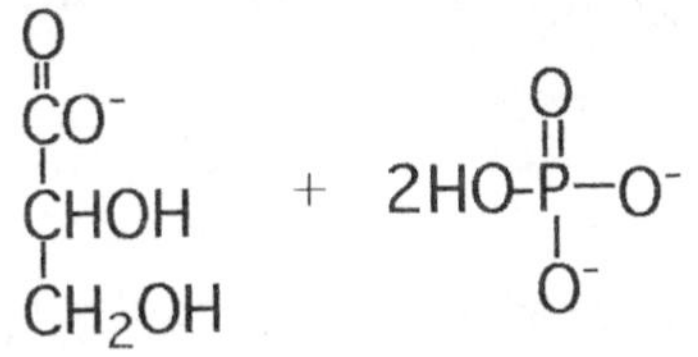

Chapter 19 Carbohydrates

<u>19.1</u> Following are Fischer projections for the four 2-ketopentoses. They consist of two pairs of enantiomers.

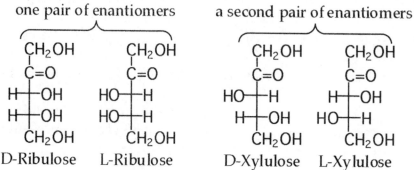

<u>19.2</u> D-Mannose differs in configuration from D-glucose only at carbon 2. One way to arrive at the structures of the α and β forms of D-mannopyranose is to draw the corresponding α and β forms of D-glucopyranose, and then invert the configuration in each at carbon 2.

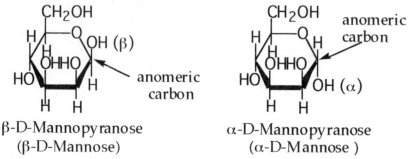

<u>19.3</u> D-Mannose differs in configuration from D-glucose only at carbon 2.

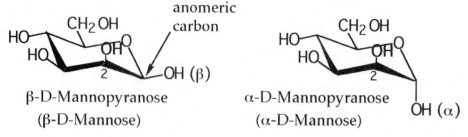

<u>19.4</u> Following is a Haworth projection and a chair conformation for this glycoside.

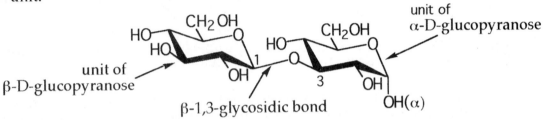

19.5 The β-glycosidic bond is between carbon 1 of the left unit and carbon 3 of the right unit.

19.7 Carbohydrates make up about three-fourths of the dry weight of plants. Less than one percent of the body weight of humans is made up of carbohydrates.

19.9 The most abundant D-aldohexose in the biological world is glucose.

19.11 D-glucose

19.13 The designations D and L refer to the configuration of the stereocenter farthest from the aldehyde or ketone group on a monosaccharide chain. When the monosaccharide chain is drawn as a Fischer projection, a D-monosaccharide has the -OH on this carbon on the right; an L monosaccharide has it on the left.

19.15 In D-glucose, carbons 2, 3, 4, and 5 are stereocenters. In D-ribose, carbons 2, 3, and 4 are stereocenters. D-Glucose is one of 16 possible stereoisomers; D-ribose is one of eight possible stereoisomers.

19.17 First draw the D form of each, and then its mirror image.

19.19 Each carbon of a monosaccharide and a disaccharide has an oxygen that is able to participate in hydrogen bonding with water molecules.

19.21 An anomeric carbon is the hemiacetal or acetal carbon in the cyclic form of a monosaccharide. Said another way, it is the carbon that was the carbonyl carbon in the open-chain form of the monosaccharide.

19.23 The designation β means that the -OH group on the anomeric carbon of a cyclic hemiacetal is on the same side of the ring as the terminal -CH$_2$OH group. The designation α means that it is on the opposite side of the ring from the terminal - CH$_2$OH group.

19.25 No. The hydroxyl groups on carbon 2, 3, and 4 of α-D-glucose are equatorial, but the hydroxyl group on carbon 1 is axial.

19.27 First compare each Haworth projection with that of β-D-glucose. Compound (a) differs in configuration at carbon 3. Compound (b) differs in configuration at carbons 2 and 3.

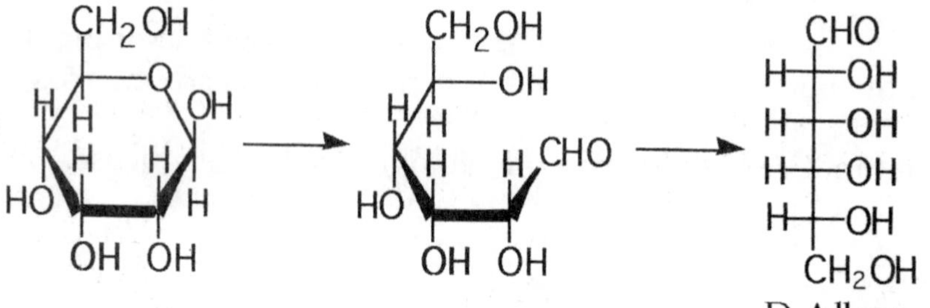

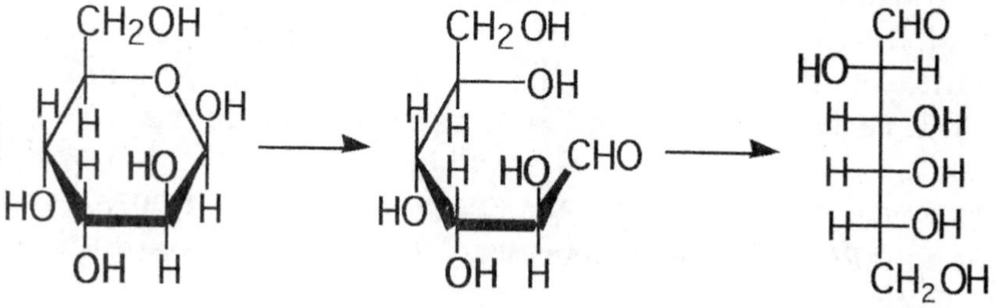

19.29 During mutarotation, the α and β forms of a carbohydrate are converted to an equilibrium mixture of the two. Mutarotation can be detected by observing the change in optical activity over time as the two forms equilibrate.

19.31 The specific rotation of α-L-glucose changes to -52.7°.

19.33 A glycosidic bond is the bond to the -OR group of the cyclic acetal form of any monosaccharide. A glucosidic bond is the bond to the -OR group of a cyclic acetal form of glucose.

19.35 Each aldehyde group is reduced by $NaBH_4$ to a primary alcohol.

(a)
```
      CH₂OH
   H ─┼─ OH
  HO ─┼─ H
  HO ─┼─ H
   H ─┼─ OH
      CH₂OH
```

(b)
```
      CH₂OH
   H ─┼─ OH
   H ─┼─ OH
   H ─┼─ OH
      CH₂OH
```

19.37 In reduction of the ketone group of D-fructose, the -OH group at carbon 2 may be either on the right or the left in the Fischer projection.

```
      CH₂OH                    CH₂OH              CH₂OH
      C=O                   HO ─┼─ H           H ─┼─ OH
  HO ─┼─ H      NaBH₄      HO ─┼─ H     +     HO ─┼─ H
   H ─┼─ OH   ────────►     H ─┼─ OH           H ─┼─ OH
   H ─┼─ OH                  H ─┼─ OH           H ─┼─ OH
      CH₂OH                    CH₂OH              CH₂OH

    D-Fructose              D-Mannitol         D-Sorbitol
```

19.39 Review Section 17.8 on keto-enol tautomerism and your answer to Problem 17.67. The intermediate in this conversion is an enediol; that is, it contains a carbon-carbon double bond with two OH groups on it.

```
      CHO                     H-C-OH                  CH₂OH
   H ─┼─ OH                    ‖                      C=O
  HO ─┼─ H                   C-OH                 HO ─┼─ H
   H ─┼─ OH        ⇌     HO ─┼─ H       ⇌          H ─┼─ OH
   H ─┼─ OH               H ─┼─ OH                  H ─┼─ OH
      CH₂OPO₃²⁻            H ─┼─ OH                     CH₂OPO₃²⁻
                             CH₂OPO₃²⁻

    D-Glucose               An enediol             D-Fructose
    6-phosphate             intermediate           6-phosphate
```

19.41 Sucrose contains one unit of D-glucose and one unit of D-fructose. Lactose contains one unit of D-galactose and one unit of D-glucose. Maltose contains two units of D-glucose.

19.43 The anomeric carbons of both monosaccharide units in sucrose are involved in forming the glycosidic bond. In maltose and lactose, one of the monosaccharide units is in the cyclic hemiacetal form, which is in equilibrium with the open chain form and, therefore, undergoes oxidation.

19.45 (a) Both monosaccharides are units of D-glucose. (b) The glycosidic bond can be described as α-1,1- or β-1,1- depending on which D-glucose unit you take as the reference. (c) Trehalose is not a reducing sugar, because both anomeric carbons participate in forming the glycosidic bond. (d) Trehalose will not undergo mutarotation for the reason given in (c).

19.47 Cellulose, starch, and glycogen are all composed of units of D-glucose. The glycosidic bonds are alpha in starch and glycogen, and beta in cellulose.

19.49 Glycogen is stored in roughly equal amounts in the liver and muscle tissue.

19.51 In their digestive systems, cattle have micororganisms that contain beta-glucosidases. These enzymes catalyze the hydrolysis of the glycosidic bonds of cellulose. We do not have these enzymes in our digestive systems.

19.53 One way to construct each repeating disaccharide is to draw a disaccharide of two units of D-glucose, and then modify each glucose unit to make it the appropriate unit in alginic acid or pectic acid.

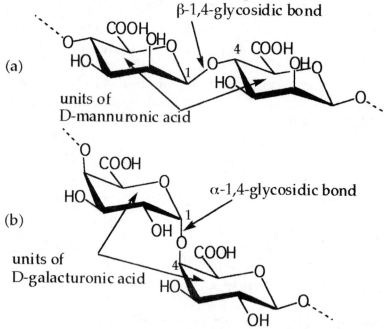

19.55 (a) The negative charges are provided by -OSO$_3^-$ and -COO$^-$ groups. (b) The higher the degree of polymerization, the better the anticoagulant activity.

19.57 The difference is two hydrogens. L-Ascorbic acid is the reduced form of this vitamin; L-dehydroascorbic acid is its oxidized form. The L indicates that each has the L configuration at carbon 5.

19.59 It is used to monitor blood glucose levels in diabetics and pre-diabetics.

Chapter 19 Carbohydrates

19.61 (a) L-Fucose is an L-aldohexose. (b) What is unusual about it in human biochemistry is that it belongs to the L series of monosaccharides, and it has no oxygen atom at carbon 6. (c) If its terminal -CH₃ group were converted to a -CH₂OH group, the monosaccharide formed would be L-galactose.

19.63 High-fructose corn syrup, as its name suggests, is produced from corn syrup by partial enzyme-catalyzed hydrolysis of cornstarch to D-glucose and then partial enzyme-catalyzed isomerization of D-glucose to D-fructose.

19.65 During boiling in water with an acid catalyst, some of the sucrose is hydrolyzed to glucose and fructose, and fructose is sweeter than sucrose.

19.67 (a) The left unit is D-glucuronic acid, which is derived from D-glucose. The right unit is a sulfate ester derived from N-acetyl-D-galactosamine, which is in turn derived from D-galactosamine. (b) The two units are joined by a β-1,3-glycosidic bond.

Chapter 20 Lipids

20.1 (a) This complex lipid is produced from the triol glycerol. Two of the hydroxyl groups are esterified with fatty acids, the third with phosphoric acid. Therefore, the lipid is broadly classified as a glycerophospholipid. The phosphate group is esterified with the hydroxyl group of the amino acid serine; therefore, the lipid belongs to the subgroup cephalins.
(b) The components present: glycerol, myristic acid, linoleic acid, phosphate, and serine.

20.3 The term hydrophobic means water fearing. Hydrophobic molecules, also called nonpolar molecules, are usually not soluble in water, but they are soluble in nonpolar organic solvents such as pentane or dichloromethane. The hydrophobic nature of lipids is important because for cells to function properly, they require separation of aqueous components (organelles). These components are surrounded by membranes that are composed of hydrophobic lipids and thus do not allow free flow of water and many other cellular molecules. Biochemical processes may be separated this way.

20.5 The melting points of fatty acids (Table 20.1) are dependent on the length of the carbon chains and the number and types of double bonds. Cis double bonds cause a kink in the carbon chain (see Section 20.2) that disrupts the London dispersion forces between the chains and lowers the melting point compared to the saturated acid, 18:0. The trans acid, 18:1, resembles a saturated acid and does not disrupt the chain as much as the cis acid, thus the melting point would be predicted to be higher. Actual melting points: 18:0 (70° C); 18:1 trans (45° C); 18:1 cis (16° C).

20.7 The diglycerides with the highest melting points will be those with two stearic acids (a saturated fatty acid). The lowest melting point will be the one with two oleic acids (a monounsaturated fatty acid).

20.9 A triglyceride containing only stearic acid (18:0) will have a higher melting point than a triglyceride containing only lauric acid (12:0). The correct answer is (b).

20.11 The highest melting triglyceride would be (a), containing palmitic and stearic (both saturated fatty acids). The lowest melting triglyceride is (c) because all fatty acids are unsaturated. Triglyceride (b), which has one unsaturated fatty acid and two saturated acids, would have a melting point in between triglycerides (a) and (c).

20.13 In order of increasing solubility in water, they are (a) < (b) < (c). Solubility in water is dependent on the presence of polar functional groups like the hydroxyl group. The greater the number of hydroxyl groups, the greater the solubility. A monoglyceride has two hydroxyl groups, a diglyceride has one hydroxyl group, and a triglyceride has none.

Chapter 20 Lipids

Fatty acids are essentially insoluble in water.

20.15 Saponification of the triglyceride hydrolyzes the ester bonds yielding glycerol and the sodium salts of the fatty acids palmitic, stearic, and linoleic.

20.17 (a) The fluidity of a membrane is dependent on the types of fatty acids attached to the hydroxyl groups on carbons 1 and 2 of the glycerophospholipid. This region is often called the nonpolar tail of the lipid. Unsaturated fatty acids esterified to the glycerol will cause the membrane to be more fluid than if the acids are saturated.
(b) The surface polarity of a membrane is dependent on the alcohol molecule (inositol) and the phosphate group. This is called the polar head of the lipid.

20.19 Most lipids in membranes are the complex lipids: phospholipids (glycerophospholipids and glycolipids) and sphingolipids. Also present is the steroid cholesterol.

20.21 The difference among the four glycerophospholipids is the nature of the polar group (choline, ethanolamine, serine, inositol). Inositol, with five hydroxyl groups that can hydrogen bond with water, has the greatest polarity.

20.23 All three of the glycerophospholipids have at least one polar group that enhances water solubility. However, phosphatidyl serine has three, a positively-charged nitrogen, a negatively-charged carboxyl, and phosphate that can associate with water molecules by ion-dipole interaction.

20.25 No, there is not random distribution of complex lipids in membranes. For example, in the red blood cells, lecithins are more abundant on the outside and cephalins are more abundant on the inside.

20.27 The presence of cholesterol in a membrane is expected to contribute to the fluidity of the membrane by disrupting the London forces between the carbon chains of the fatty acids. It disrupts interactions similarly to a cis double bond, thereby raising fluidity.

20.29 (a) The cholesterol molecule has eight stereocenters (see structure below).
(b) Eight stereocenters will lead to 2 to the 8th power or 256 stereoisomers.
(c) Only one stereoisomer is found in nature.

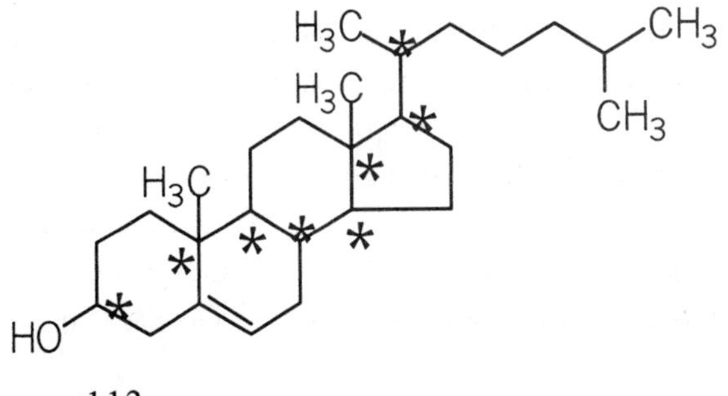

20.31 On the surface of the LDL clusters are phospholipids, free cholesterol, and proteins all of which contain polar groups that can form hydrogen bonds with water thus making the LDL clusters soluble.

20.33 The drug lovastatin inhibits the action of the enzyme, HMG-CoA reductase, which is a key enzyme in the biosynthesis of cholesterol. Thus, less cholesterol is synthesized and available to cause atherosclerotic plaques.

20.35 High density lipoproteins, called the carriers of "good cholesterol", remove cholesterol from extrahepatic tissue and deliver it to the liver. HDL circulating in the plasma picks up cholesterol perhaps by extraction from cell surface membranes and transformation to cholesteryl esters. The liver may take up HDL using a very specific HDL receptor.

20.37 Cortisol is the most important glucocorticoid hormone. It increases the production of glucose and glycogen in the liver. The carbohydrates are produced from fatty acids and amino acids that are transported to the liver. Cortisol is also an anti-inflammatory agent.

20.39 Estradiol and testosterone have similar structures based on the four-fused steroid rings. However, estradiol has a phenol group (an OH on a phenyl ring). Testosterone has two methyl branches (estradiol only one), and a carbon-carbon double bond and ketone.

20.41 (a) Progesterone and RU486 both have the standard four-fused ring system of the steroids with a ketone functional group at carbon 3. They both have methyl branches at carbon 13 and oxygen-containing functional groups directly or indirectly attached to ring D (hydroxyl for RU486 and acetyl for progesterone).
(b) They differ in that RU486 has the p-aminophenyl group at carbon 11 and a propyne group at carbon 17. Progesterone does not have these groups.

20.43 Taurocholate has several functional groups that enhance water solubility: three hydroxyl groups (carbons 3, 7, and 12), and an ionic sulfate group on the side chain.

20.45 (a) Arachidonic acid is a noncyclic fatty acid with four carbon-carbon double bonds. PGE_2 is a diunsaturated fatty acid containing several functional groups: a five-membered ring, a ketone group, and two hydroxyl groups.
(b) PGE_2 has a ketone functional group at carbon 9, whereas $PGF_{2\alpha}$ has a hydroxyl group at carbon 9.

20.47 The COX-2 enzyme catalyzes the formation of prostaglandins from arachidonic acid, but it functions only when tissue is injured; for example, in response to inflammation.

20.49 Rancidity occurs when fats and oils are exposed to air. This process is caused by reaction of oxygen with allylic hydrogens leading to cleavage of the carbon-carbon double bonds and formation of small aldehydes that have unpleasant odors and tastes. Rancidity may be inhibited by storing lipids at lowered temperatures and in dark bottles that reduce UV light, an oxidation catalyst.

20.51 Soaps are sodium or potassium salts of long chain fatty acids. Detergents are sodium salts of, usually synthetic, long chain sulfonic acids. Both soaps and detergents have an anionic hydrophilic end; soaps have a carboxylate group and detergents a sulfate group.

20.53 The anion transporter of the erythrocyte membrane exchanges chloride and bicarbonate ions (Chemical Connections 20D). The protein transporter forms a hydrophilic channel through the membrane. Protein polar groups in the channel wall facilitate the transport of hydrated chloride and bicarbonate ions. The hydrophobic groups on the outer surfaces of the helical section interact with the membrane.

20.55 (a) Nerve axons are surrounded by a lipid layer, called the myelin sheath, which is composed primarily of sphingomyelin. The protective lipid layer provides insulation and enhances the rapid conduction of electrical signals.
(b) In multiple sclerosis, there is slow degradation of the protective myelin sheath. This exposes the nerve axons to damage and causes afflicted individuals to lose strength and coordination.

20.57 The glycolipid that accumulates in Fabry's disease contains the monosaccharides galactose and glucose.

20.59 Progesterone is secreted during the menstrual cycle after ovulation (see Figure 20.7). Its purpose is to prevent another ovulation if the egg is fertilized. Contraceptive agents give the user a level of progesterone that sends the message that there is a fertilized egg; therefore, ovulation does not occur.

20.61 Indomethacin, a non-steroidal anti-inflammatory agent, acts by inhibiting the cyclo-oxygenase enzymes that catalyze the formation of prostaglandins from arachidonic acid. Prostaglandins cause an inflammatory response so inhibiting their formation will reduce inflammation.

20.63 The leukotrienes are synthesized from arachidonic acid as are prostaglandins, but not through the same pathway using the COX enzymes. Thus, the NSAIDS inhibit the cyclo-oxygenases and prostaglandin synthesis, but do not effect leukotriene synthesis. Aspirin and other NSAIDS agents can reduce inflammation, but they do not alter the physiological actions of the leukotrienes.

20.65 (Study membrane structure in Figure 20.2). Three regions of the membrane need to be considered in order to define transport: 1) the cytoplasmic surfaces and 2) the exterior surfaces that are comprised primarily of the polar heads of complex lipids (purple balls) and 3) the membrane interior composed of the hydrophobic tails of the lipids. To diffuse, molecules must pass through all three regions, two polar regions and one nonpolar region. Polar molecules interact favorably with the surface regions by hydrogen bonding and ion-dipole interactions, but they repel the hydrophobic barrier, thus they would not diffuse through the membrane. In addition, polar molecules would prefer to interact with water that is present at each surface. On the other hand, small, nonpolar molecules are able to diffuse because they slip through surface regions containing mainly hydrophobic components (lipid patches).

20.67 Although prostaglandins and thromboxanes have different structures, the physiological production of both is slowed by COX inhibitors. This suggests that there must be a common step in their syntheses. Prostaglandins are synthesized from arachidonic acid using the COX enzymes. Thromboxanes are synthesized from the prostaglandin PGH.

20.69 Coated pits are areas on a cell membrane that have high concentrations of LDL receptors and participate in transporting LDLs into the cell.

20.71 Detergents, like sodium dodecyl sulfate and taurocholate, have several common features. They both have a polar (anionic) sulfate functional group on one end and nonpolar, hydrophobic regions (carbon-hydrogen skeleton) on the other end.

20.73 The LDL receptor recognizes a specific type of protein, apoB-100, present on the surface of the LDL.

20.75 Aspirin and most other NSAIDS inhibit both COX enzymes thus leading to the undesirable side effects. Celebrex, the new anti-inflammatory drug, inhibits only COX-2.

20.77 Both cortisone and the synthetic drug, prednisolone, are used to treat inflammatory diseases. They have several similar structural features: they both have the four-fused steroid ring system with methyl branches, common substituents on ring D, and a carbon-carbon double bond/ketone in ring A. Differences: Prednisolone has an extra carbon-carbon double bond in ring A and a hydroxyl group in ring C. Cortisone has a ketone functional group in ring C.

Chapter 21 Proteins

21.1 The dipeptide Val-Phe:

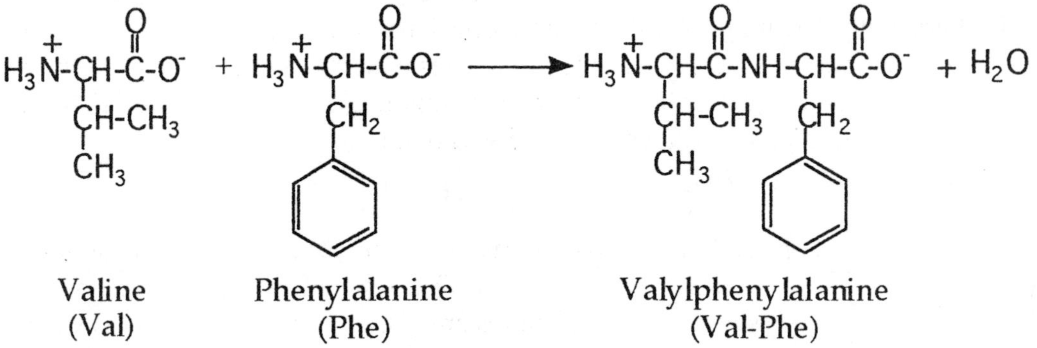

Valine   Phenylalanine   Valylphenylalanine
(Val)   (Phe)   (Val-Phe)

21.2 Salt bridges

21.3 (a) Ovalbumin is a protein found in eggs that serves to store nutrients for the embryo.
(b) Myosin is essential for muscle contraction.

21.4 Proteins are classified broadly into two groups: those that are soluble in water (globular proteins), and those that are insoluble in water (fibrous proteins). The fibrous proteins, because they are water-insoluble, can serve a structural role in organisms.

21.5 See Table 21.1. Tyr has a hydroxyl group on the phenyl ring; Phe has a hydrogen at the same position.

21.7 Arg has the highest percentage of nitrogen at 32%. The four nitrogens have a weight of $4 \times 14 = 56$. The molecular weight of Arg is 174. Percent of N $= 56 \div 174 \times 100 = 32\%$. His has 27% nitrogen.

21.9 The amino acid Pro, which has a five-membered ring containing a single nitrogen atom, is classified as a pyrrolidine:

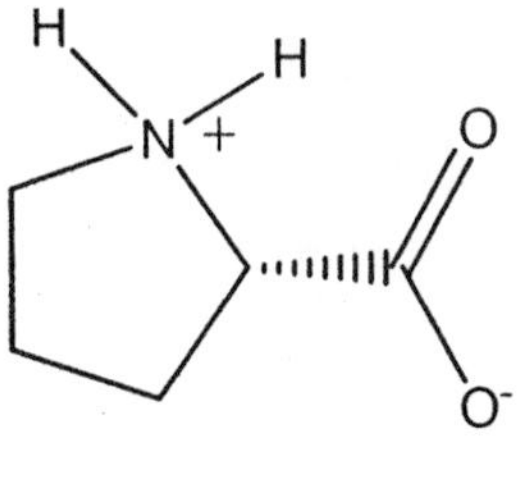

21.11 Proteins that are ingested in meat, fish, eggs, etc. are degraded to free amino acids and then used to build proteins that carry out specific functions (see Section 21.1). Two very important functions include structural integrity (collagen, keratin) to hold body tissue together, and biological catalysis to assist in cellular reactions, especially metabolism.

21.13 They both have the three-carbon skeleton found in Ala, but Phe has a phenyl group substituted on the β-methyl group:

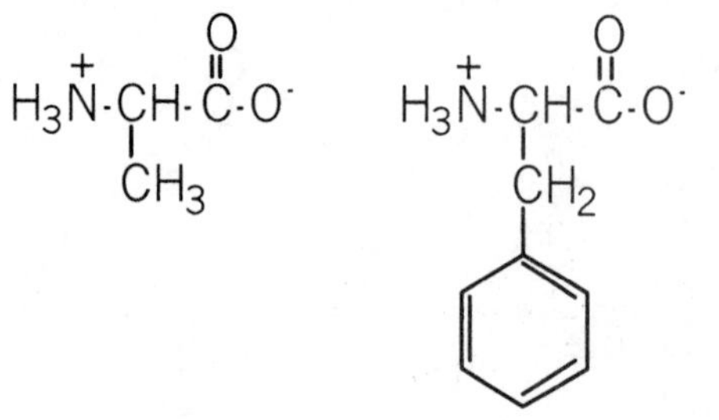

$$H_3\overset{+}{N}\text{-CH-}\overset{O}{\overset{\|}{C}}\text{-}O^- \qquad H_3\overset{+}{N}\text{-CH-}\overset{O}{\overset{\|}{C}}\text{-}O^-$$

Alanine
(Ala, A)

Phenylalanine
(Phe, F)

21.15 Structures of L-Val and D-Val:

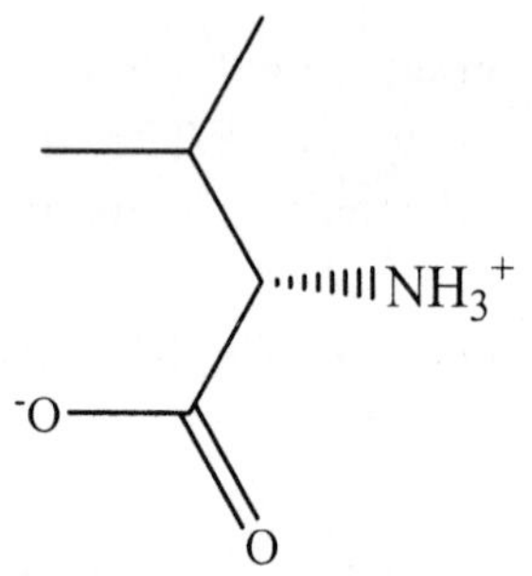

D-valine

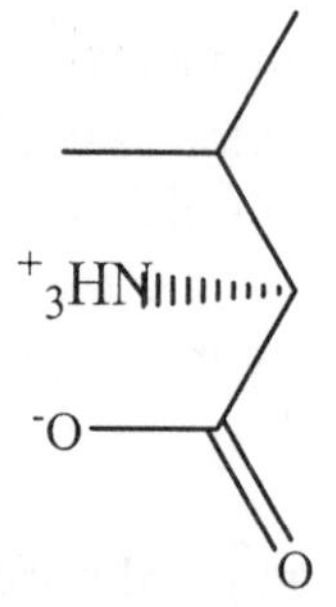

L-valine

21.17 Ala at its isoelectric point has the following structure:

$$CH_3\text{-}\underset{NH_3^+}{CH}\text{-}COO^-$$

Ala acting as a base to neutralize an acid:

$$CH_3\text{-}\underset{NH_3^+}{CH}\text{-}COO^- + H_3O^+ \longrightarrow CH_3\text{-}\underset{NH_3^+}{CH}\text{-}COOH + H_2O$$

Ala acting as an acid to neutralize a base:

$$CH_3\text{-}\underset{NH_3^+}{CH}\text{-}COO^- + OH^- \longrightarrow CH_3\text{-}\underset{NH_2}{CH}\text{-}COO^- + H_2O$$

21.19 The pK_a values for Val are approximately 2 for the carboxyl group and 9 for the amino group. The structures at pH 1 and 12 are:

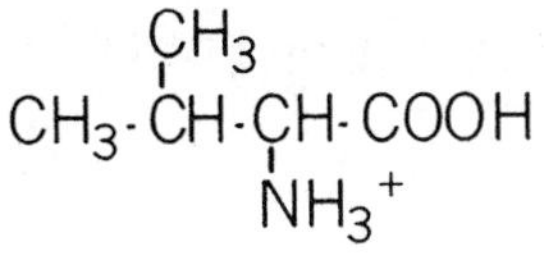
CH3
CH3-CH-CH-COOH
NH3+

at pH 1

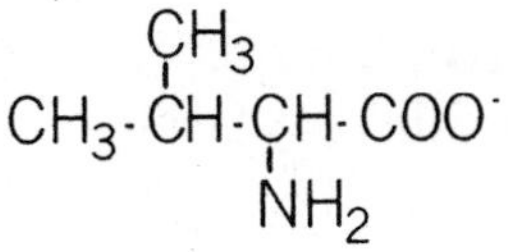
CH3
CH3-CH-CH-COO-
NH2

at pH 12

21.21 The tripeptide Thr-Arg-Met:

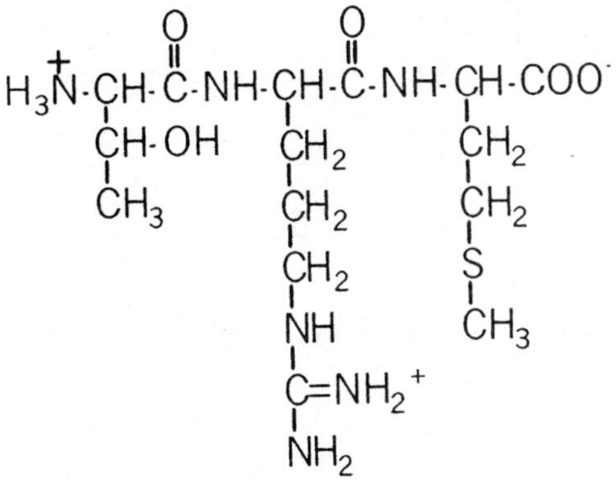
O
O
H3N+-CH-C-NH-CH-C-NH-CH-COO-
CH-OH
CH3
CH2
CH2
CH2
NH
C=NH2+
NH2
CH2
CH2
S
CH3

21.23 The dipeptide Leu-Pro:

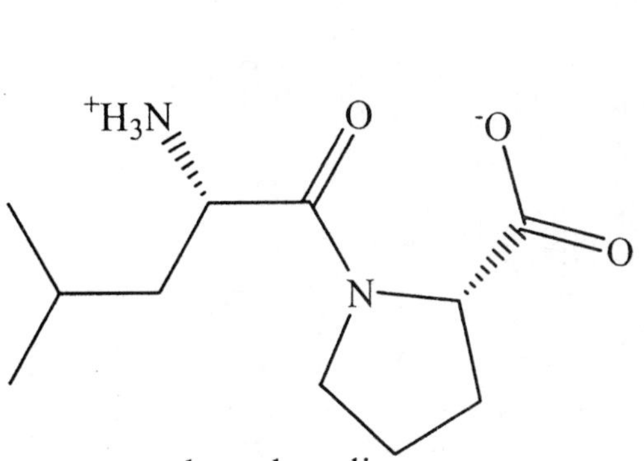
+H3N
O
-O
O
N

leucyl-proline

21.25 Structures of Ala-Gln and Gln-Ala:

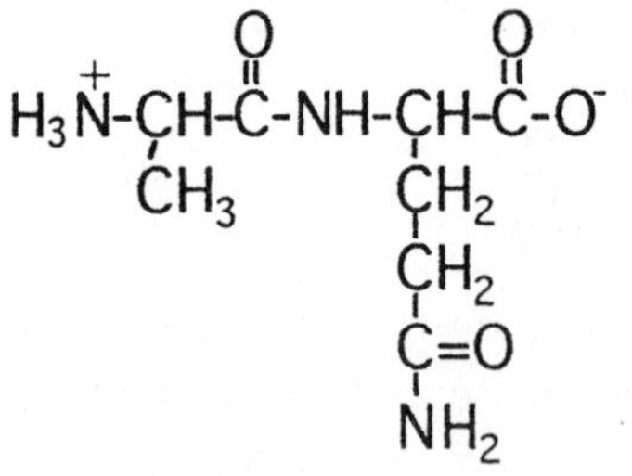

Alanylglutamine
(Ala-Gln)

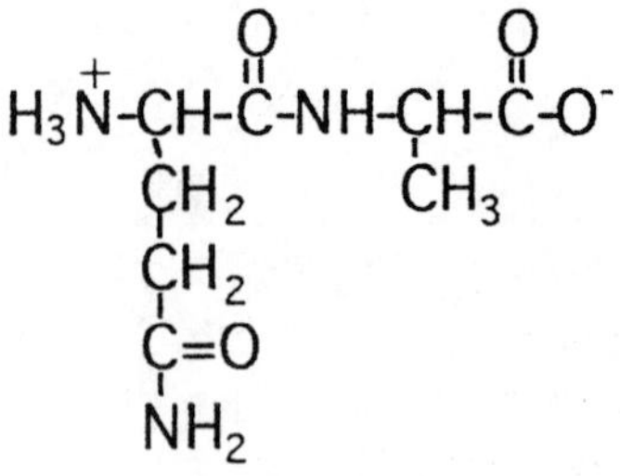

Glutaminylalanine
(Gln-Ala)

21.27 (a) Structure of Met-Ser-Cys:

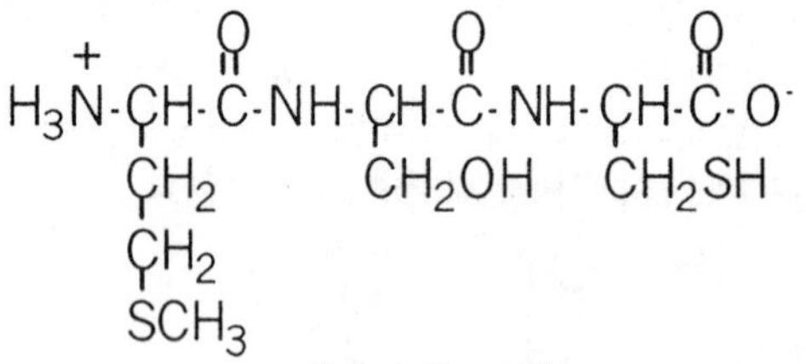

Met-Ser-Cys

(b) The pK_a of the carboxyl group is about 2, so at pH = 2:

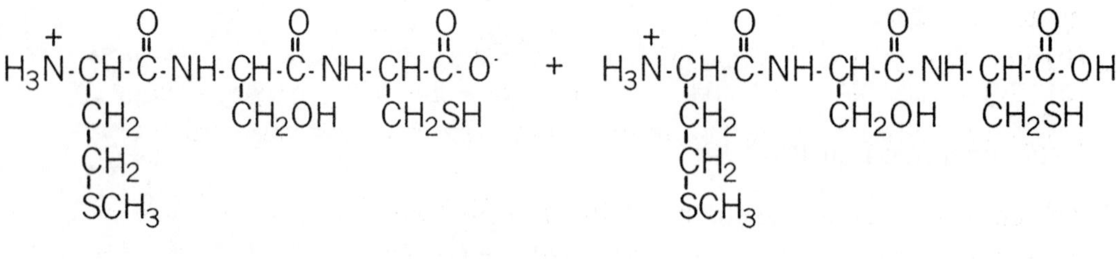

The pK_a of the α-amino group is about 9:

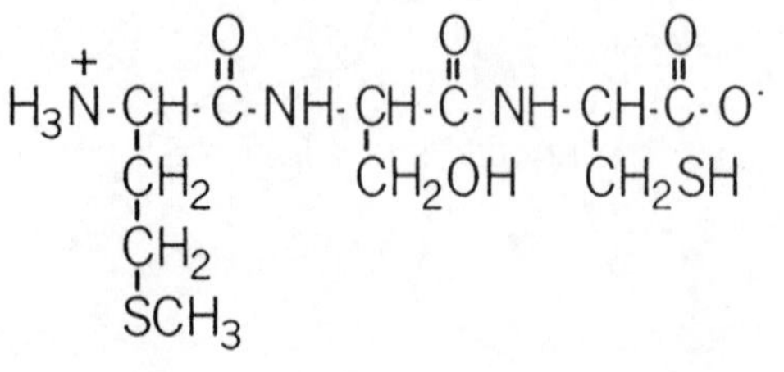

At pH 7.0 net charge = 0

$$H_2N \cdot CH \cdot \overset{\overset{\displaystyle O}{\|}}{C} \cdot NH \cdot CH \cdot \overset{\overset{\displaystyle O}{\|}}{C} \cdot NH \cdot CH \cdot \overset{\overset{\displaystyle O}{\|}}{C} \cdot O^-$$

with side chains:

CH₂ — CH₂OH — CH₂SH
CH₂
SCH₃

At pH 10, net charge = -1

21.29 The precipitated protein at its isoelectric point has an equal number of positive and negative charges (no net charge). The protein molecules clump together and form aggregates. Adding acid protonates the carboxylate groups on amino acid residue side chains and at the C-terminus giving the protein a net positive charge due to the protonated nitrogen atoms of the amino groups. The charged protein molecules now repel each other and are soluble.

21.31 (a) The number of different tetrapeptides may be calculated by assuming that any of the four amino acids may be located in any of the positions. Therefore, there are 4 to the power of 4 or 256 possible tetrapeptides.
(b) If all 20 amino acids are available, the total number of possibilities would be 20 to the 4th power or 160,000.

21.33 Leu, which has a side chain made up of only hydrocarbon, is a nonpolar amino acid. The fewest changes in the protein would result if the Leu were substituted by other nonpolar amino acids like Val or Ile.

21.35 (a) Tropocollagen, the triple helix form of collagen, is considered to be a form of quaternary structure.
(b) The collagen fibril form is defined by quaternary structure.
(c) Collagen fibers are considered to be an example of quaternary structure.
(d) The repeating sequence of amino acids refers to the primary structure.

21.37 The carboxylic acid side chain of Glu has a pK_a of 4.25. At low pH values, the side chain carboxyl groups are protonated and thus have no charge. At higher pH (above 4.25), the side chains begin to be deprotonated and become negatively charged. When the polypeptide wraps into an α-helix, this brings the side chains in close proximity (Figure 21.6a). Above a pH of 4.25, the negatively charged side chains interact unfavorably (repel each other) which destabilizes the helix, and the polypeptide forms a random coil. When the side chains are protonated at low pH, the neutral side chains allow for the formation of an alpha helix.

21.39 Box 1: the carboxyl terminus of the top polypeptide.
Box 2: the amino terminus of the lower polypeptide.

Box 3: a region of antiparallel β-pleated sheet between the top and bottom polypeptides.
Box 4: a section of random coil structure.
Box 5: a region of hydrophobic bonding.
Box 6: an intrachain disulfide bond between Cys residues.
Box 7: a region of α-helix.
Box 8: a salt bridge (ionic interaction) between the side chains of Asp and Lys.
Box 9: a hydrogen bond that is part of the tertiary structure.

21.41 The urea disrupts hydrogen bonding between protein backbone C=O and H-N groups.

21.43 In integral membrane proteins, nonpolar side chains of proteins move to the outside surface so they may interact favorably (form hydrophobic bonds) with the nonpolar lipids in the membrane bilayer.

21.45 Cytochrome c is an example of a conjugated protein, one that contains, in addition to the protein portion, a non-amino acid portion called a prosthetic group. In cytochrome c the prosthetic group is an iron porphyrin, also called a heme. The two parts are held together by noncovalent interactions. Cytochrome c is a conjugated protein.

21.47 Structure of a saccharide linked to the amide nitrogen of an Asn residue:

21.49 Collagen is a glycoprotein.

21.51 The principal protein in hair is keratin which has many Cys residues that are linked via disulfide bonds (-S-S-). These strong covalent bonds are important in holding the shape of hair. To straighten or change the hair style, it is treated with a reducing agent that breaks some of the disulfide bonds. The hair is then set into the new, desired shape and treated with an oxidizing agent to reform disulfide bonds thus preserving the new, attractive hair style.

21.53 Wiping the skin with 70% alcohol (70% alcohol: 30% water), kills bacteria thus sterilizing the skin before the injection. Ethanol at a concentration of 70% penetrates the cell membrane of bacteria and denatures their proteins.

Chapter 21 Proteins

21.55 AGE refers to Advanced Glycation End products that accumulate in our bodies throughout a lifetime. These products are formed non-enzymatically when the amino groups of proteins (i.e. Lys side chains) react with aldehyde and ketone functional groups in carbohydrates. The new covalent linkage is called an imine (Chemical Connections 21B). The longer we live, the more AGE products accumulate in tissue. Because of faster, more efficient metabolism, younger people degrade and eliminate these undesirable biochemicals.

21.57 The chemical hydroxyurea, sold under the trade name Droxia, has recently been approved by the Food and Drug Administration for treatment of sickle cell anemia (Chemical Connections 21D). Hydroxyurea enhances the production of fetal hemoglobin (HbF) in the bone marrow. HbF, because of its amino acid sequence does not sickle and clog capillaries as does HbS. Although bone marrow still produces HbS, it is diluted with HbF and thus somewhat relieves the symptoms of the disease.

21.59 The major goal of proteomics is to identify and determine the function of every protein in all cells and tissues. It is hoped that such a protein profile will help physicians diagnose and treat diseases.

21.61 The fiberscope (Chemical Connections 21G) is an instrument that focuses a laser beam onto damaged tissue to perform medical procedures. When the energetic laser beam is absorbed by tissues, the proteins are heat denatured and joined together. This procedure may be used to seal damaged tissue like bleeding ulcers.

21.63 (a) Using four different amino acids in a chain length of two amino acids, the total possible is 4^2 or 16 possible dipeptides.
(b) 20^2 or 400 possible dipeptides.

21.65 The term β-barrel is used to describe the quaternary structure of integral membrane proteins. These proteins, because they transverse the membrane bilayer, have a nonpolar surface that interacts hydrophobically with the lipid bilayer. The nonpolar surface is the hydrophobic faces of the beta sheets.

21.67 During aging, the triple helices of collagen are crosslinked by the formation of covalent bonds between lysine side chains. This is a form of quaternary structure.

21.69 (a) Val and Ile, both with nonpolar side chains, interact by forming hydrophobic bonds.
(b) The side chains of Glu and Lys interact ionically to form a salt bridge.
(c) Hydrogen bonding between hydroxyl groups.
(d) Hydrophobic interactions between alkyl groups.

21.71 Gly has no stereocenter; therefore, it does not exist as enantiomers.

21.73 The amino acid Asp has three pH values: 1.88 for the α-carboxyl group; 4.25 for the side-chain carboxyl; 9.6 for the α-amino group. The forms present at pH 2:

Major form present at
pH 2.0; net charge = 0

Minor form present at
pH 2.0; net charge = +1

21.75 Carbohydrates are polar because of the presence of hydroxyl groups that can hydrogen bond with water. It is predicted that adding carbohydrates to a protein will enhance the solubility of the protein in water.

Chapter 22 Enzymes

22.1 The word catalyst is a general term used to define agents that speed up chemical reactions. Enzymes are natural catalysts, composed of proteins and sometimes RNA, that catalyze reactions in biological cells.

22.3 Lipases, enzymes that catalyze the hydrolysis of ester bonds in triglycerides, are not very specific as they work on many different triglyceride structures with about the same effectiveness. It is predicted that the two triglycerides containing palmitic acid and oleic acid would be hydrolyzed at about the same rate.

22.5 There are thousands of different kinds of biochemicals in an organism and thousands of reactions are required to synthesize and degrade the compounds. The mild conditions inside the cell are not conducive to fast reaction rates so each reaction requires an enzyme for catalysis under physiological conditions.

22.7 Hydrolases use water to break bonds in a substrate usually leading to two smaller products; for example, the enzyme acetylcholinesterase in Table 22.1. Water is always involved in a hydrolase reaction. Lyases catalyze the addition of a group to a double bond or the removal of a group to form a double bond. The added or removed group may be water, but it is not always one of the substrates. An example of a lyase is aconitase in Table 22.1.

22.9 (a) The reactant and product have the same molecular formula, but they have different molecular structures, so they are isomers. The enzyme that catalyzes the reaction is classified as an isomerase.
(b) Water is a reactant that hydrolyzes the two amide bonds in urea. The enzyme is classified as a hydrolase.
(c) Succinate is oxidized and FAD is reduced. The enzyme is classified as an oxidoreductase.
(d) Aspartase catalyzes the addition of a functional group to a double bond. The enzyme is classified as a lyase.

22.11 A cofactor is a nonprotein part of an enzyme that must be present for catalytic activity. Metal ions are often cofactors. When a cofactor is an organic molecule like FAD or heme, it is called a coenzyme.

22.13 Noncompetitive inhibitors slow enzyme activity by binding to sites other than the active site where the substrate (S) binds. A reversible inhibitor (I) binds to and dissociates from the enzyme thus setting up an equilibrium:

$$E + I \rightleftharpoons EI$$

An irreversible inhibitor binds permanently to the enzyme and alters its structure:

$$E + I \longrightarrow E\text{-}I$$

22.15 At very low concentrations of substrate, the rate of an enzyme-catalyzed reaction increases in a linear fashion with increasing substrate concentrations. At higher concentrations of substrate, the active sites of the enzyme molecules become saturated with substrate and the rate of increase is slowed. The rate is at a maximum level when all of the enzyme active sites are occupied with substrate. That maximum rate cannot be exceeded even if more substrate is added.

22.17 (a) Normal body temperature is 37° C (98.6° F) and a fever of 104° F is about 40° C. According to the curve, the bacterial enzyme is more active with the fever.
(b) The activity of the bacterial enzyme decreases if the patient's temperature decreases to 35° C.

22.19 (a) Plot of the pH dependence of pepsin activity:

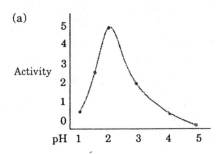

(b) Optimum pH is about 2.
(c) Pepsin is predicted to have no catalytic activity at pH 7.4.

22.21 (A) Large red circle = enzyme with active site
(B) Pentagon = glucose
(C) Smaller triangle = cadmium ion
(D) Larger triangle = magnesium ion
(E) Square = ATP
(F) Black circle = fluoroglucose
(a) ABDE
(b) ADEF
(c) ABCE

22.23 A survey of the amino acids present at enzyme active sites has shown an abundance of amino acid residues with acidic and basic side chains. The most prominent amino acids include His, Asp, Arg, and Glu. The properties of these amino acids lead one to the conclusion that acid-base chemistry (proton transfer) is an important chemical process during enzyme action.

22.25 In competitive inhibition, the maximum reaction rate achieved is the same with or without an inhibitor (Figure 22.11). However, in the presence of a competitive inhibitor, a higher substrate concentration is required to reach the maximum rate. The maximum rate for a noncompetitive inhibitor is always lower at any substrate concentration and can never equal the rate of the uninhibited reaction.

22.27 At first, one might expect the inhibition of phosphorylase action by caffeine to be a case of traditional noncompetitive inhibition because the inhibitor apparently does not bind at the active site. However, glycogen phosphorylase is known to be an allosteric enzyme and the classical terms competitive and noncompetitive are not used to define inhibition. The term negative modulator is used to define the action of caffeine on the allosteric enzyme, phosphorylase.

22.29 The terms zymogen and proenzyme are both used to define an inactive form of an enzyme.

22.31 The action of a protein kinase is to catalyze the transfer of a phosphate group from ATP to another molecule, in this case a tyrosyl residue of an enzyme:

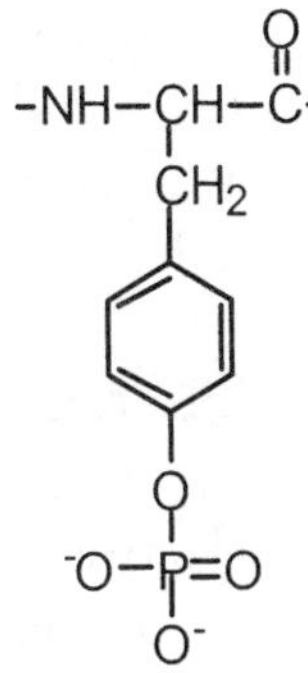

$$-NH-CH-C-$$

22.33 Glycogen phosphorylase b is converted to glycogen phosphorylase a by the action of an enzyme called phosphorylase kinase. The kinase transfers phosphate groups from ATP to phosphorylase b. Two ATPs are required because the phosphorylase has two subunits, each of which must be phosphorylated:

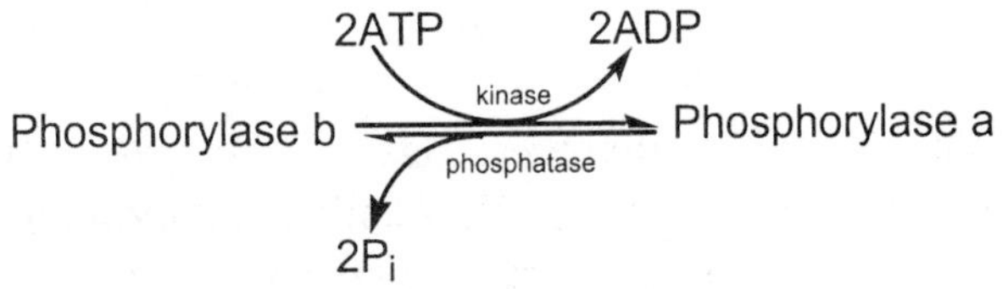

22.35 The reaction in Section 27.9 describes the transfer of an amino group from alanine to α-ketoglutarate to produce pyruvate and glutamate. Although enzymes catalyze reactions in both directions, one direction is usually predominant under physiological conditions. The new name, ALT, defines the predominant direction of the reaction and the source of the amino group being transferred.

22.37 Lactate dehydrogenase, which is tetrameric, has two different kinds of subunits called H and M. Thus, LDH exists as isoenzymes. The subunit that dominates in heart cells is the H type. Therefore, the isoform, H_4, is monitored in the diagnosis of a heart attack.

22.39 Patients who have had duodenal ulcer surgery are administered digestive enzymes that may be in short supply after the surgical procedure. The enzyme preparation contains proteases that hydrolyze dietary proteins and lipases that hydrolyze dietary fats.

22.41 Succinylcholine has a chemical structure similar to that of acetylcholine so they both can bind to the acetylcholine receptor of the muscle end plate. The binding of either choline causes a muscle contraction. However, the enzyme acetylcholinesterase hydrolyzes succinylcholine only very slowly compared to acetylcholine. Muscle contraction does not occur as long as succinylcholine is still present and acts as a relaxant.

22.43 The bacterium that causes stomach ulcers, *H. pylori*, protects itself from stomach acid using the enzyme urease. This enzyme catalyzes the hydrolysis of urea forming ammonia, a strong base. The chemical process continually bathes the bacterial cell with ammonia that neutralizes the stomach acid.

22.45 Antibiotics like amoxillin and tetracycline are used to treat ulcers caused by *H. pylori*.

22.47 In the enzyme pyruvate kinase, the $=CH_2$ of the substrate phosphoenolpyruvate sits in a hydrophobic pocket formed by the amino acids Ala, Gly, and Thr. The methyl group on the side chain of Thr rather than the hydroxyl group is in the pocket. Hydrophobic interactions are at work here to hold the substrate into the active site.

22.49 Humans are not able to synthesize folic acid, but must obtain it in the diet or from intestinal microorganisms. Early sulfa drugs like sulfanilamide were toxic to humans, but their toxicity was not due to inhibition of folic acid synthesis. Newer sulfa drugs are much less toxic to humans.

22.51 All of the ACE inhibitors shown in Figure 22F have the following common functional groups: a five- or six-membered ring containing a nitrogen atom; a carboxylate group substituted on the ring; an amide group; and two negative charges on each molecule.

Chapter 22 Enzymes

22.53 Enzymes from thermophilic microorganisms that live in deep-sea vents or Yellowstone National Park geysers have enzymes that are active at high temperatures.

22.55 (a) Vegetables such as green beans, corn, and tomatoes are heated to kill microorganisms before they are preserved by canning. Milk is preserved by the heating process, Pasteurization.
(b) Pickles and sauerkraut are preserved by storage in vinegar (acetic acid).

22.57 The amino acids removed from the decapeptide are His and Leu.

22.59 The zinc ion is called a metal cofactor.

22.61 The enzyme alanine aminotransferase is elevated in patients with hepatitis.

22.63 No, the inhibition by nerve gas is not an example of competitive inhibition. Competitive inhibition is a reversible process, whereas the action of a nerve gas is irreversible. Adding more substrate does not remove the covalently-bound nerve gas from the enzyme.

22.65 The enzyme that catalyzes the reaction in Section 28.5 moves an amino group from one molecule (glutamate) to another (pyruvate). The enzyme is a transferase.

22.67 A ribozyme is a form of RNA that has catalytic activity; that is, it functions as an enzyme. Ribozymes are involved in splicing reactions (self cleavage) in post-transcriptional processing. The enzyme peptidyl transferase, responsible for protein synthesis (Section 25.5), has recently been identified as a ribozyme.

Chapter 23 Chemical Communications: Neurotransmitters and Hormones

23.1 G protein is a membrane-bound enzyme that hydrolyzes GTP to GDP or GMP.

23.3 A chemical messenger is a biomolecule such as a hormone that binds to a receptor in a cell membrane with the purpose of initiating some change inside the cell. A secondary messenger is a biomolecule like cAMP that transmits the action of the chemical messenger to the inside of the cell, usually with amplification.

23.5 The role of calcium ions in the release of neurotransmitters is most easily understood when one considers a nerve impulse reaching the presynaptic site (Chemical Connection 23A). This causes the opening of ion channels and the entry of calcium ions into the nerve cell that leads to the release of neurotransmitter molecules into the synapse.

23.7 Lactation is controlled by hormones from the pituitary gland. The anterior lobe makes prolactin that stimulates the growth of the mammary gland. The posterior lobe of the pituitary produces the hormone oxytocin, which stimulates milk flow.

23.9 The neurotransmitter acetylcholine is stored in presynaptic vesicles (Chemical Connection 23A). When a nerve impulse arrives, ion channels open and allow calcium ions to enter the nerve cell. The vesicles then fuse with the membrane and release acetylcholine into the synapse where the neurotransmitter molecules bind to receptors on the postsynaptic side. Binding of acetylcholine to the receptors opens ion channels. The flow of ions creates electrical signals.

23.11 Cobratoxin causes paralysis by acting as a nerve system antagonist—it blocks the receptor and interrupts the communication between neuron and muscle cell. The botulin toxin prevents the release of acetylcholine from the presynaptic vesicles.

23.13 Taurine has an amino group and an acidic group, hence it may be classified as an amino acid. However, it differs from the amino acids found in proteins. Taurine has a sulfonic acid group instead of a carboxylic acid and the functional groups are on different carbon atoms than in protein amino acids.

23.15 γ-Aminobutyric acid (GABA) is also an amino acid; however, the two functional groups are not bound to the same carbon atom as in the case of protein amino acids. Another name for GABA is 4-aminobutanoic acid.

23.17 (a) Two monoamine neurotransmitters in Table 23.1 are dopamine and serotonin.
(b) They both act by binding to adrenergic receptors, thus continuing the nerve impulse from one neuron to another.

(c) The drug Deprenyl may be used to increase dopamine levels. Serotonin levels are increased by the drug Prozac.

23.19 Protein kinase is activated by interaction with cyclic AMP which dissociates the regulatory and catalytic subunits. The catalytic subunit is then activated by phosphate transfer from ATP (Figure 23.4).

23.21 Product from the MAO-catalyzed oxidation of dopamine:

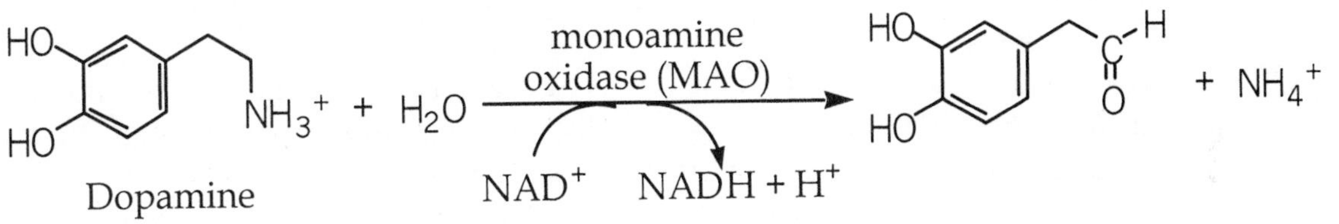

23.23 (a) Amphetamines stimulate adrenergic neurotransmission.
(b) Reserpine decreases the concentration of adrenergic hormones thus causing a sedative effect.

23.25 Removal of the methyl amino group from epinephrine gives a product containing the aldehyde functional group (Section 23.5F).

23.27 (a) The ion channel is blocked by the ion-translocating protein (see Figure 23.4).
(b) When the blocking protein is phosphorylated, it moves to open the channel.
(c) Cyclic AMP activates the protein kinase that phosphorylates the ion-translocating protein.

23.29 The brain peptides called enkephalins are by chemical nature pentapeptides.

23.31 A kinase catalyzes the phosphorylation of inositol-1,4-diphosphate to inositol-1,4,5-triphosphate:

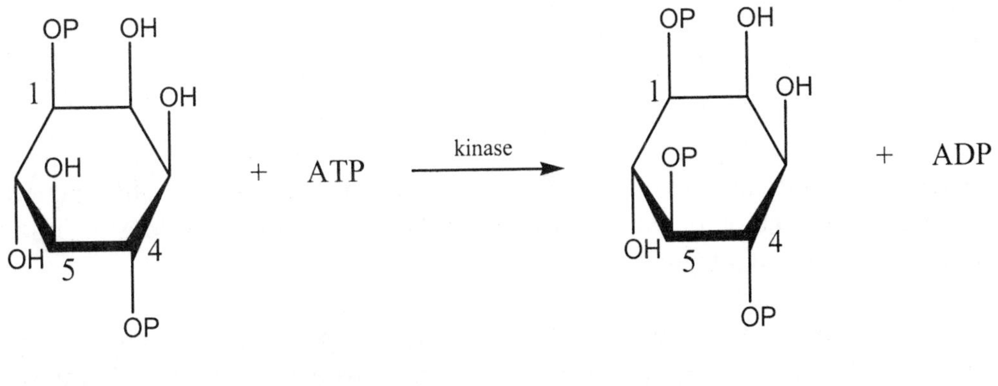

$P = -PO_3^{2-}$

23.33 Calcium sparks or puffs are sudden influxes of calcium ions from extracellular sources or from the endoplasmic reticulum. Calcium waves are slow fluctations (increases and decreases) of calcium ion concentration inside the cell.

23.35 The calcium ion concentration must increase 100 to 250 times normal in order to achieve fusion.

23.37 The botulin toxin, found in *C. botulinum,* prevents the release of acetylcholine from the presynaptic vesicle. No neurotransmitter reaches the receptor molecules.

23.39 The neurofibrillar tangles found in the brains of Alzheimer's patients are composed of tau proteins. Mutated tau proteins, which normally interact with the cytoskeleton, grow into these tangles instead, thus altering normal cell structure.

23.41 Drugs that increase the concentration of the neurotransmitter acetylcholine may be effective in the treatment of Alzheimer's disease. A chemical found in Chinese herb tea inhibits the enzyme acetylcholinesterase. This chemical agent should increase the concentration of the deficient neurotransmitter.

23.43 Cocaine binds to the dopamine transporter, like a reversible inhibitor, preventing the reuptake of dopamine. The dopamine is not transported back to the original neuron and stays in the synapse, increasing the continuous firing of signals.

23.45 Parkinson's disease is characterized by a reduced number of dopamine neurons. The latest treatment of the disease involves the transplantation of human embryonic dopamine neurons into the patient's brain. These new neurons indeed produce dopamine thus relieving most of the symptoms of the disease.

23.47 The enzyme nitric oxide synthase catalyzes the formation of NO from arginine.

23.49 The sulfonyl urea moiety found in drugs for diabetes:

$$\text{C}_6\text{H}_5-\overset{\overset{\displaystyle O}{\|}}{\underset{\underset{\displaystyle O}{\|}}{\text{S}}}-\text{NH}-\overset{\overset{\displaystyle O}{\|}}{\text{C}}-\text{NH}-\text{CH}_2\text{CH}_2\text{CH}_2-$$

23.51 Insulin acts by binding to receptors on the surface of fat cells. This binding process triggers the formation of cyclic GMP inside the cell and this enhances the transport of glucose into the adipocyte.

23.53 Tamoxifen binds to estradiol receptors on the nuclei of cells and prevents the binding of the normal hormone, estradiol. When estradiol binds to the receptor, it initiates processes that often stimulate growth of breast cancer cells.

23.55 During nerve transmission, sodium and potassium ions flow through the membrane. This transport process sets up a chemical gradient to continue the nerve impulse. The ion-translocating protein controls the flow of the ions through the membrane (Figure 23.4).

23.57 Yes, the two peptides endorphin and enkephalins have some identical amino acids and their structures are similar enough so they can bind to the same receptor.

23.59 G proteins are usually bound to the inner side of cell membranes and usually close to receptors for neurotransmitters.

23.61 (a) The hormones vasopressin and dopamine have little in common except that they both elicit their actions by binding to specific receptor proteins.
(b) Vasopressin binds to a receptor that activates protein kinase C and involves the second messenger phosphatidyl inositol. Dopamine is a neurotransmitter that is amplified by cyclic AMP secondary messenger.

23.63 Cholera toxin permanently activates the G protein which leads to overproduction of cyclic AMP and continuously opens ion channels. This results in the constant outflow of ions and accompanying water of hydration causing dehydration and diarrhea.

23.65 Glucose levels in the serum would decrease as the sugar is rapidly taken up by cells.

23.67 Luteinizing hormone, a peptide, acts through signal transduction involving G protein and cyclic AMP. Progesterone, a steroid hormone, acts by stimulating transcription of specific genes directly without the aid of secondary messengers, so that particular proteins are produced.

Chapter 24 Nucleotides, Nucleic Acids, and Heredity

<u>24.1</u> Structure of UMP:

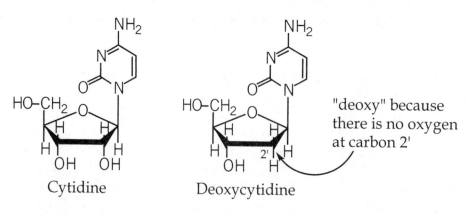

<u>24.3</u> Scientists and physicians have identified hundreds of heredity diseases. Sickle cell anemia is one you have studied in detail (Chemical Connections 21D).

<u>24.5</u> (a) Nucleotide components: a nitrogen base (AGTC or AGTU); a sugar (ribose or deoxyribose); a phosphate group esterified at the 5' position of the sugar.
(b) Nucleoside components: a nitrogen base and a sugar.

<u>24.7</u> See Figure 24.1. Thymine and uracil are both based on the pyrimidine ring. However, thymine has a methyl substituent at carbon 5 whereas uracil has a H. All of the other ring substituents are the same.

<u>24.9</u> (a) Structure of cytidine: (b) Structure of deoxycytidine:

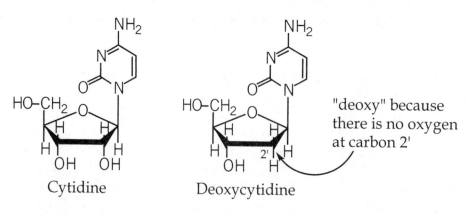

<u>24.11</u> D-ribose and 2-deoxy-D-ribose have the same structure except at carbon 2. D-ribose has a hydroxyl group and hydrogen on carbon 2, whereas deoxyribose has two hydrogens (see Section 24.2B):

Chapter 24 Nucleotides, Nucleic Acids, and Heredity

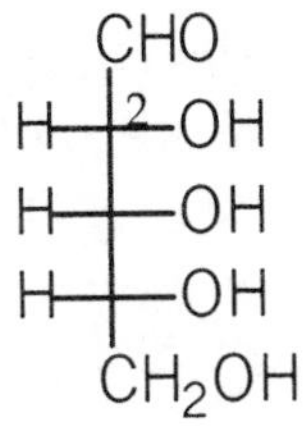

D-Ribose

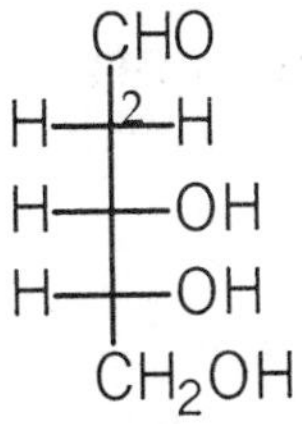

2-Deoxy-D-ribose

24.13 In RNA, carbons 3' and 5' of the ribose are linked by ester bonds to phosphates. Carbon 1 is linked to the nitrogen base with an N-glycosidic bond.

24.15 (a) Structure of UMP; see Problem 24.1.
(b) Structure of dAMP:

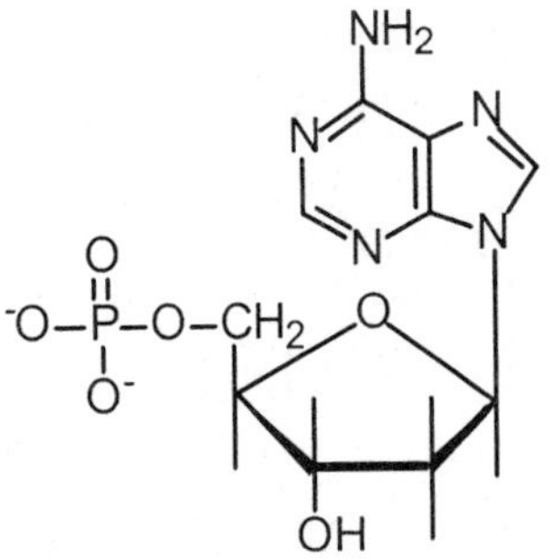

24.17 (a) One end will have a free 5' phosphate or hydroxyl group that is not in phosphodiester linkage. That end is called the 5' end. The other end, the 3' end, will have a 3' free phosphate or hydroxyl group.
(b) By convention the end drawn to the left is the 5' end. A is the 5' end and C is the 3' end.

24.19 Two hydrogen bonds form between uracil and adenine. This base pairing is the same as with adenine and thymine shown in Figure 24.5.

24.21 Histones are proteins with a high content of two basic amino acids, Lys and Arg. Recall that these two amino acid residues in proteins will have positively-charged side chains at physiological pH. On the other hand, DNA at physiological pH will have many negatively charged groups due to the ionized phosphates in the backbone. These two types of molecules will form very strong electrostatic interactions or salt bridges.

24.23 The superstructure of chromosomes is comprised of many elements. DNA and histones combine to form nucleosomes that are wound into chromatin fibers. These fibers are further twisted into loops and minibands to form the chromosome superstructure (see Figure 24.8).

24.25 tRNA molecules range in size from 73 to 93 nucleotides per chain. mRNA molecules are larger with an average of 750 nucleotides per chain. rRNA molecules can be as large as 3000 nucleotides.

24.27 All three types of RNA will have a sequence complementary to a portion of a DNA molecule because the RNA is transcribed from a gene.

24.29 Ribozymes, or catalytic forms of RNA, are involved in post-transcriptional splicing reactions that hydrolyze larger RNA molecules into smaller more active forms. For example, tRNA molecules are formed in this way.

24.31 Messenger RNA immediately after transcription contains both introns and exons. The introns are hydrolyzed out by the action of ribozymes that catalyze splicing reactions on the RNA.

24.33 Satellite DNA, in which short nucleotide sequences are repeated hundreds and even thousands of times, are not expressed into proteins. They are found at the ends and centers of chromosomes and are necessary for stability.

24.35 See Figure 24.5. Functional groups that serve as hydrogen bond acceptors: carbonyl groups; ring nitrogens. Hydrogen bond donors: primary amines bonded to the rings; secondary amines in the rings.

24.37 See Figure 24.5. Moving from the top of the molecules down, a carbonyl group on G accepts a hydrogen bond from a primary amine of C. An N-H group on G donates a hydrogen bond to a ring nitrogen of C. A primary amine on G donates a hydrogen bond to a carbonyl oxygen of C.

24.39 A replication fork is the point on the DNA where synthesis of a new strand begins (replication, see Figure 24.11). In human cells, the average chromosome may have several hundred forks.

24.41 The chromosome superstructure is held together partly by ionic interactions between the positive charges on histone proteins and the negative phosphates of the DNA backbone. Removing some of the positive charges on lysyl residues in the histones weakens the DNA:histone interaction. This allows some regions on the chromosome to open up.

24.43 Eukaryotic helicases are composed of six different protein subunits that form a ring with a hollow core. The single-stranded DNA that is unwound moves through the core (see Figure 24.11).

24.45 Because phosphodiester bonds are formed between the 3' and 5' positions of the ribose 2'-dATP is not expected to be a primer.

24.47 The DNA polymerases join nucleotides using phosphodiester bonds to form DNA.

24.49 The DNA polymerases catalyze the formation of 3', 5' phosphodiester bonds.

24.51 The modified G residue could be removed using the nucleotide excision repair (NER) pathway (Section 24.7).

24.53 Deamination of the base cytosine transforms it to uracil which creates a mismatch. What was a C-G pair now becomes a U-G mismatched pair that must be removed by BER.

24.55 During the excision repair process, the first step involves removal of the damaged base by hydrolysis of the N-C glycosidic bond between the sugar and the base. That leads to a free hydroxyl group on C-1' of the sugar. This is called an AP (apurinic or apyrimidinic) site because it does not have a base. Enzymes called DNA glycosylases form the AP sites.

24.57 The polymerase chain reaction (PCR) may be used to make millions of copies of selected DNA fragments. The enzyme used for the process is a heat resistant polymerase from thermophilic bacteria. A temperature of 95° C is required to unwind the DNA to prepare it for copying by the polymerase. The enzyme will, of course, function much faster at the elevated temperature compared to room temperature. Most enzymes denature at these high temperatures so the heat stable polymerase provides a special advantage.

24.59 Structure of fluorouracil:

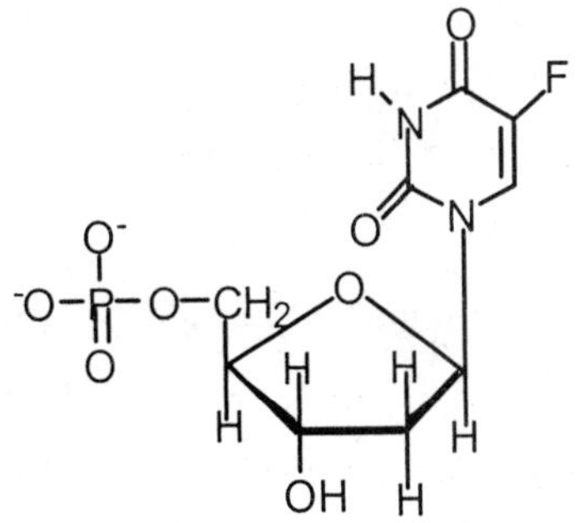

24.61 The sequence of repeated nucleotides in vertebrate telomers is TTAGGG.

24.63 In normal somatic cells that divide in a cyclic fashion throughout the life of an organism, chromosomes lose about 50 to 200 nucleotides from telomers at each cell division. The shortening of the telomers acts as a clock by which the cells count the number of times they have divided. The cells die a programmed death after a certain time period. Cancer cells have an enzyme called telomerase that is able to extend the shortened telomers by synthesis of new ends. The activity of telomerase appears to confer immortality on the cancer cells.

24.65 See Chemical Connections 24C. DNA samples from mother, child, and potential father are amplified by PCR, hydrolyzed into fragments using restriction enzymes, and the fragments separated and analyzed by electrophoresis. Figure 24C shows the results of these procedures. Lanes 2, 3, and 4 show the results for the father, child, and mother respectively. The child's DNA contains 6 bands and the mother's has five bands, all of which match those of the child, confirming the mother:child relationship. The band from the alleged father also contains six bands of which only three match the DNA of the child. A child is expected to inherit only half of his/her genes from the father so a match of three bands is sufficient to prove paternity.

24.67 See Chemical Connections 24D. Information about an individual's genetic profile will make it possible to prescribe drugs in proper dosages and to avoid adverse reactions to certain drugs and drug combinations. The major drug-metabolizing enzyme in humans is cytochrome P-450. This enzyme, which catalyzes the addition of hydroxyl groups to drugs thus making them more water soluble, occurs in different forms which have different catalytic activities. By knowing what form an individual has, it will be possible to prescribe the correct dosage of a drug so that it remains in the body long enough to be most effective.

24.69 In cell death by apoptosis, only a few scattered cells die at a time, rather than in large clusters as is found in necrosis.

24.71 All of the information necessary to produce a functional organism is present in its DNA. When an organism is conceived, the egg and sperm cells unite to form the zygote. When cells grow and divide, each one must be identical to the original. Thus, there must be a mechanism by which DNA replication occurs over and over again with only minimal error.

24.73 The nitrogen base is linked to the sugar by an N-glycosidic bond. The phosphate group is linked by an ester bond to the hydroxyl group on the 5'-carbon of the sugar.

24.75 Native DNA is the largest nucleic acid.

24.77 Mol % A = 29.3; Mol % T = 29.3; Mol % G = 20.7; Mol % C = 20.7.

Chapter 25 Gene Expression and Protein Synthesis

25.1 Transcription begins when the DNA double helix begins to unwind at a point near the gene that is to be transcribed. The superstructures of DNA (chromosomes, chromatin, etc.) break down to the nucleosome (histones plus DNA). Binding proteins interact with the nucleosomes making the DNA less dense and more accessible. The helicase enzymes then begin to unwind the double helix.

25.2 Codons for His: CAU; CAC. Anticodons: GTA; GTG.

25.3 Valine + tRNA (specific for Val) + ATP

25.4 Sticky ends after cleavage of DNA by a restriction enzyme:

—CCT   CGATTG—
—GGAGC   TAAC—

25.5 Answer (c); gene expression refers to both processes, transcription and translation.

25.7 Protein translation occurs on the ribosomes.

25.9 Helicases are enzymes that catalyze the unwinding of the DNA double helix prior to translation. The helicases break the hydrogen bonds between base pairs.

25.11 The termination signal for transcription is at the 5' end of the DNA.

25.13 The "guanine cap" methyl group is located on nitrogen number 7 of guanine.

25.15 A codon, the three-nucleotide sequence that specifies amino acids for protein synthesis, is located on an mRNA molecule.

25.17 (a) According to Example 25.2, the anticodon is CGA.
(b) The amino acid is alanine.

25.19 All organisms use the same genetic code; that is, the sequence of three bases codes for the same amino acid. This observation strongly suggests that all living organisms have the same origin, a common ancestor.

25.21 The ribosomal sites participating in translation are called A, P, and E.
Site A: the aminoacyl site where the incoming amino acid-tRNA binds.
Site P: the peptidyl site where the growing peptide is bound before the new amino acid

is added.
Site E: the exit site where the tRNA, after release of its amino acid, is bound just before it leaves.

25.23 Elongation factors are proteins that participate in the process of tRNA binding and movement of the ribosome on the mRNA during the elongation process in translation.

25.25 Enhancers are regulatory regions on the DNA that may be thousands of nucleotides away from the transcription site. They may be brought into the vicinity of transcription by special loop-forming proteins called sensors (see Figure 25.9).

25.27 Proteosomes are cylindrical assemblies of a number of protein subunits with protease activity. Proteosomes play a role in post-translational degradation of damaged proteins. Proteins damaged by age or proteins that have misfolded are degraded by the proteosomes.

25.29 (a) Silent mutation: assume the DNA sequence is TAT which will code for UAU on the mRNA. Tyrosine is incorporated into the protein. Now assume a mutation in the DNA to TAC. This will pair to UAC in mRNA. Again, the amino acid will be tyrosine.
(b) Lethal mutation: original DNA sequence is CTT, which transcribes into GAA on mRNA. This codes for the amino acid glutamic acid. The DNA mutation ATT will lead to UAA, a stop signal that incorporates no amino acid.

25.31 Yes, a harmful mutation may be carried as a recessive gene from generation to generation with no individual demonstrating symptoms of the disease. Only when both parents carry recessive genes does an offspring have a 25% chance of inheriting the disease.

25.33 Restriction endonucleases are enzymes that recognize specific sequences on DNA and catalyze the hydrolysis of phosphodiester bonds in that region thus cleaving both strands of the DNA (see Section 25.8). These enzymes are useful for the preparation of recombinant DNA.

25.35 Mutation by natural selection is an exceedingly long, slow process that has occurred for centuries. Each natural change in the gene has been ecologically tested and found usually to have a positive effect or the organism is not viable. Genetic engineering, where a DNA mutation is done very fast, does not provide sufficient time to observe all of the possible biological and ecological consequences of the change.

25.37 Vitravene, an antisense drug used by AIDS patients, is made up of nucleotides, but it has a backbone different from the normal phosphodiester linkages. It has a sulfur atom substituted for an oxygen atom in each phosphate group (see Figure 25A). This functional group change slows the physiological degradation of the drug.

<u>25.39</u> The AIDS virus recognizes a protein called CD4, which is on the helper T cells, part of the immune system.

<u>25.41</u> The coding region of the HIV-1 protease was linked to the promoter of αA-crystallin, a protein found mainly in the lens of the eye. The DNA sequence containing the gene for HIV-1 protease and the promoter were transfected into a mouse. The result was transgenic mice that became blind soon after birth and developed cataracts. These mice may be used to test AIDS drugs. If a new drug is active against the protease, the onset of cataracts will be delayed (Chemical Connections 25D).

<u>25.43</u> A single mutation on a gene—a change from G to T—transforms the EGF proto-oncogen to an oncogen which causes a form of human bladder carcinoma.

<u>25.45</u> The tumor suppressor protein, p53, promotes DNA repair. When DNA is damaged (x-rays, chemicals, etc.), p53 protein concentration increases. p53 controls the cell cycle; it maintains the cycle between cell division and DNA replication. The cell thus gains time in repairing the DNA damage.

<u>25.47</u> (a) Mandatory base pairing in tRNA: between nucleotides near the 5' end and the 3' end. These interactions help hold the tRNA in the familiar tertiary structure.
(b) Absence of base pairing: in the anticodon loop.

<u>25.49</u> Enzymes called aminoacyl tRNA synthetases are responsible for bringing together the proper pair. The enzymes match the amino acid and one of the correct tRNA molecules carrying the right anticodon.

<u>25.51</u> Degenerate, when used to define the genetic code, means that there is more than one codon that can specify the same amino acid. For example, UUU and UUC both code for Phe.

<u>25.53</u> The third base is not irrelevant in codons for Phe, Leu, Ile, Met, Tyr, His, Gln, Asn, Lys, Asp, Glu, Cys, Trp, Ser, and Arg. For 15 of the 20 amino acids in proteins, there is a codon combination where the third base is not irrelevant. Of the 16 possible first two base combinations, in eight of them, the third base is irrelevant (CU, GU, UC, CC, AC, GC, CG, and GG).

<u>25.55</u> The new endonuclease would be of no value in producing recombinant DNA. There are many, many G-C pairs in DNA, and cleavage at every one of these would lead to hundreds of very small fragments some as small as a single nucleotide.

Chapter 26 Bioenergetics. How the Body Converts Food to Energy

26.1 Chemical energy present in food molecules is extracted and converted to a usable form by the process of catabolism. The energy derived from the degradation of different types of molecules is collected in the form of the energy-rich molecule, ATP.

26.3 (a) Mitochondria have two membranes, a highly-folded inner membrane, and an outer membrane (see Figure 26.3).
(b) The outer membrane is permeable to the diffusion of small ions and molecules. Special transport processes are required to move molecules through the inner membrane.

26.5 Cristae are the folds that are present in the inner mitochondrial membrane. The folds provide extensive surface area for the concentration of enzymes and other components required for metabolism.

26.7 Each ATP molecule has two phosphate anhydride bonds that release a substantial amount of energy when they are hydrolyzed during metabolic processes (see Figure 26.5):

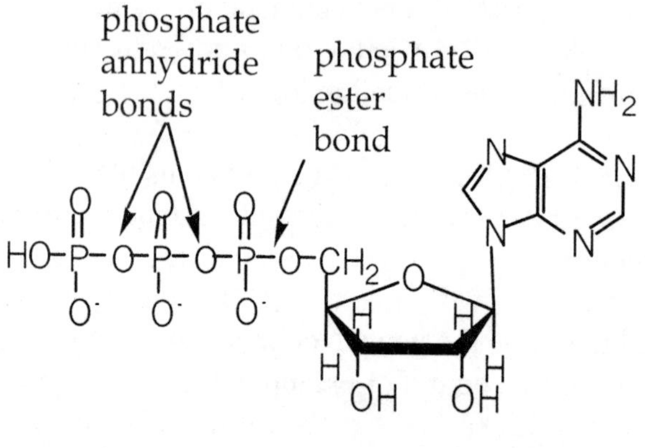

26.9 In reactions (a) and (b) the same type of anhydride bond is hydrolyzed. When the reactions are measured under standard conditions, the energy yield is about the same, 7.3 kcal/mole. In the cell, however, the concentration of ADP is very low, compared to ATP, and ADP is rarely used for energy.

26.11 The chemical bond between the ribitol and phosphate in FAD is a phosphate ester (see Figure 26.6). It is formed when a phosphate reacts with the hydroxyl (alcohol) group on the sugar.

26.13 Two nitrogen atoms in the flavin ring that are linked to carbon (N=C) are reduced to yield $FADH_2$.

26.15 (a) The most important carrier of phosphate groups is ATP.
(b) The most important carriers of hydrogen ions and electrons from redox reactions are NADH and $FADH_2$.
(c) Coenzyme A carries acetyl groups.

26.17 An amide linkage is formed to bring together the amine group of mercaptoethylamine and the carboxyl group of pantothenic acid (Figure 26.7).

26.19 No, the reactive part of CoA is the thiol group (-SH) at the end of the molecule.

26.21 Most fats and carbohydrates are degraded in catabolism to the compound, acetyl CoA.

26.23 α-Ketoglutarate is the only C-5 intermediate in the citric acid cycle.

26.25 FAD is used in the citric acid cycle as a coenzyme to oxidize succinate to fumarate.

26.27 Fumarase is classified as a lyase because it catalyzes the addition of water to a double bond.

26.29 ATP is not produced directly by the citric acid cycle. GTP, a reactive molecule with the same amount of energy as ATP, is produced in Step 5 and may be used for some energy-requiring processes.

26.31 The stepwise degradation and oxidation of acetyl CoA in the citric acid cycle is very efficient in the extraction and collection of energy. Rather than occurring in one single burst, the energy is released in small increments and collected in the form of reduced cofactors, NADH and $FADH_2$.

26.33 Carbon-carbon double bonds are present in the citric acid cycle intermediates fumarate and cis-aconitate.

26.35 When α-ketoglutarate is oxidized in Step 4 of the citric acid cycle, the electrons are transferred to NAD^+ to make the reduced form, NADH.

26.37 The mobile carriers of electrons in the electron transport chain are coenzyme Q and cytochrome c.

26.39 The proton translocator ATPase is a complex that resembles a rotor engine. It rotates every time a proton passes through the inner membrane (Figure 26.10).

26.41 During oxidative phosphorylation water is formed from protons, electrons, and oxygen on the matrix side of the inner membrane. This occurs when electrons are shuttled through complex IV (Figure 26.10).

26.43 (a) For each pair of protons translocated through the ATPase complex, one molecule of ATP is generated. Each pair of electrons that enters oxidative phosphorylation at complex I yields three ATP.
(b) Each C-2 fragment, which represents the carbons in acetyl CoA, yields 12 ATP.

26.45 Protons are translocated through a proton channel formed by the Fo part of the ATPase which has 12 protein subunits embedded in the inner membrane (Figure 26.10).

26.47 The catalytic unit of ATPase is composed of α and β subunits (Figure 26.10). This part of the ATPase catalyzes the formation of ATP:

$$ADP + P_i \longrightarrow ATP + H_2O$$

26.49 The energy, in kcal, from the oxidation of 1 g of acetate by the citric acid cycle:
Molecular wt. acetate = 59 g/mole. 1 g of acetate = $1 \div 59$ = 0.017 mole.
Each mole of acetate produces 12 moles of ATP (see Problem 26.43(b)).
0.017 mole x 12 = 0.204 mole of ATP.
0.204 mole of ATP x 7.3 kcal/mole = 1.5 kcal.

26.51 (a) Muscle contraction takes place when thick protein filaments called myosin slide past thin protein filaments called actin. The hydrolysis of ATP by myosin, an ATPase, drives the alternating association and dissociation of actin and myosin. This causes the contraction and relaxation of muscles (see Section 26.8).
(b) The energy in muscle contraction comes from ATP.

26.53 Glycogen phosphorylase is activated by phosphorylation of serine residues in the protein subunits. The phosphoryl groups are transferred from ATP (see Problem 22.33).

26.55 Iron-sulfur clusters, active as electron carriers in electron transport, are composed of iron and sulfur ions. Both of these elements are toxic to cells so it is important to carefully disguise the toxicity. Iron-sulfur clusters are present in two predominant types, [2Fe-2S] and [4Fe-4S], and serve as prosthetic groups buried within proteins. This makes the ions less accessible to toxic reactions.

26.57 The antibiotic oligomycin inhibits the catalytic subunits of ATPase, thus it stops phosphorylation of ADP. Although it acts as an effective antibiotic, its toxic effect on ATPase does not allow its use in humans.

Chapter 26 Bioenergetics. How the Body Converts Food to Energy

26.59 (a) Cytochrome P-450 is an enzyme that catalyzes the hydroxylation of various natural and synthetic substrates. The hydroxyl group is derived from molecular oxygen.
(b) The source of the reactant oxygen is the air we breathe. Most (90%) of the oxygen we breathe is used for respiration.

26.61 Number of g of acetic acid that must be metabolized to yield 87.6 kcal of energy:
87.6 kcal ÷ 7.3 kcal/mole ATP = 12 moles of ATP. 12 moles of ATP are released from the oxidation of 1 mole of acetate. 1 mole of acetate (MW = 60) = 60 g of acetic acid.

26.63 The mechanical energy generated from the translocation of protons in oxidative phosphorylation is first displayed in the rotating ion channel of ATPase.

26.65 Citrate and malate, intermediates in the citric acid cycle, both have carboxyl groups and a hydroxyl group.

26.67 Myosin, the thick filament in muscle, is an enzyme that acts as an ATPase.

26.69 Isocitrate has two stereocenters.

26.71 The proton channel is located in the Fo unit of ATPase that is embedded in the inner membrane.

26.73 All the sources of energy used for ATP synthesis are not completely known at this time. Most of the energy comes from proton translocation. Some energy for proton pumping comes also from breaking the covalent bond of oxygen (reduction of oxygen to water).

26.75 Succinate dehydrogenase catalyzes the oxidation of succinate to fumarate. FAD becomes reduced in the reaction and transfers its electrons directly to the electron transport chain at complex II.

Chapter 27 Specific Catabolic Pathways: Carbohydrate, Lipid, and Protein Metabolism

27.1 (a) Using the ATP yield data in Table 27.2, stearic acid releases 8.1 ATP per carbon atom. (b) Lauric acid yields 7.9 ATP per carbon atom.

27.3 The major use of amino acids is in the synthesis of proteins. Proteins from ingested food are hydrolyzed and the amino acids are used to rebuild proteins that the body constantly degrades. We cannot store amino acids so we need a constant supply in our diet.

27.5 The step referred to is # 4, the aldolase-catalyzed cleavage of fructose 1,6-bisphosphate to glyceraldehyde 3-phosphate and dihydroxyacetone phosphate. Glyceraldehyde 3-phosphate metabolism continues immediately in glycolysis (Step # 5), but dihydroxyacetone phosphate must first be isomerized to glyceraldehyde 3-phosphate by an isomerase. Glyceraldehyde 3-phosphate and dihydroxyacetone phosphate are in equilibrium and removal of glyceraldehyde 3-phosphate by glycolysis drives the isomerization reaction.

27.7 (a) The steps in glycolysis that need ATP are # 1, phosphorylation of glucose and # 3, the phosphorylation of fructose 6-phosphate. (b) The steps that yield ATP are # 6, catalyzed by phosphoglycerate kinase and # 9, catalyzed by pyruvate kinase.

27.9 ATP is a negative modulator for the allosteric, regulatory enzyme phosphofructokinase, Step # 3.

27.11 The oxidation of glucose 6-phosphate by the pentose phosphate pathway produces NADPH. This reduced cofactor is necessary for many biosynthetic pathways, but especially for the synthesis of fatty acids.

27.13 The anaerobic degradation of a mole of glucose leads to two moles of lactate. Therefore, three moles of glucose produce six moles of lactate.

27.15 Using the data in Table 27.1, a net yield of 12 ATPs is produced by the glycolysis of glucose (glucose to pyruvate). Most of the ATP from glucose degradation comes from oxidation of the reduced cofactors, NADH and $FADH_2$, linked to respiration.

27.17 (a) Fructose catabolism by glycolysis in the liver yields two ATPs just like glucose. (b) Glycolytic breakdown of fructose in muscle also yields two ATPs per fructose.

27.19 Enzymes that catalyze the phosphorylation of substrates using ATP are called kinases. Therefore, the enzyme that transforms glycerol to glycerol 1-phosphate is called glycerol kinase.

27.21 (a) The enzymes are thiokinase and thiolase.
(b) "Thio" refers to the presence of the element sulfur.
(c) Both of these enzymes use Coenzyme A that contains a reactive thiol group, SH, as a substrate.

27.23 Each turn of fatty acid β oxidation yields one C-2 fragment (acetyl CoA), one $FADH_2$, and one NADH. Therefore, the total yield from three turns is three acetyl CoA, three $FADH_2$, and three NADH. There is still a portion of lauric acid left, the six-carbon compound, hexanoyl CoA.

27.25 Using the data from Table 27.2, the yield from the oxidation of one mole of myristic acid is 112 moles of ATP. The process requires six turns of β oxidation and produces 7 moles of acetyl CoA.

27.27 Under normal conditions, the body preferentially uses glucose as an energy source. When a person is well fed (balance of carbohydrates and fats and proteins), fatty acid oxidation is slowed and the acids are linked to glycerol and are stored in fat cells for use in times of special need. Fatty acid oxidation becomes important when glucose supplies begin to be depleted, for example, during extensive physical exercise or fasting or starvation.

27.29 The transformation of acetoacetate to β-hydroxybutyrate is a redox reaction using the cofactor, NADH. Acetone is produced by the spontaneous decarboxylation of acetoacetate.

27.31 Oxaloacetate produced from the carboxylation of PEP normally enters the citric acid cycle at Step 1. As we will learn in the next chapter, oxaloacetate may also be used to synthesize glucose.

27.33 Oxidative deamination of alanine:

$$CH_3-\underset{\underset{NH_3^+}{|}}{CH}-COO^- + NAD^+ + H_2O \longrightarrow CH_3-\underset{\underset{O}{\|}}{C}-COO^- + NADH + H^+ + NH_4^+$$

Alanine   Pyruvate

27.35 A glucogenic amino acid is one that may be used for the biosynthesis of glucose. These amino acids are deaminated to carbon skeletons that are degraded to pyruvate and other intermediates, and then converted to glucose by reverse glycolysis (Figure 27.9).

27.37 The compound fumarate is an intermediate in both the citric acid cycle and the urea cycle.

27.39 Most of the toxic ammonium ion is removed from the body by conversion to urea, but it may also be detoxified by reductive amidation, which is the reverse reaction of oxidative deamination and by the ATP dependent amidation of glutamate to yield glutamine (Section 27.8).

27.41 During initial hemoglobin catabolism, the heme group and globin proteins are separated. The globins are hydrolyzed to free amino acids that are recycled and the iron is removed from the porphyrin ring and saved in iron-storage protein, ferritin, for later use.

27.43 During exercise, normal glucose catabolism shifts to a greater production of lactate rather than conversion of pyruvate to acetyl CoA and entry into the citric acid cycle. This shift in metabolism is a result of a depletion of oxygen supplies. A build-up of lactate in muscle leads to a lowering of pH which effects myosin and actin action.

27.45 The acidic nature of the ketone bodies lowers blood pH. This increase in proton concentration is neutralized by the bicarbonate/carbonic acid buffer system presence in blood (Section 8.11D).

27.47 Lipomics refers to the detailed analysis of lipids in blood, cells and tissues. An ether extract of gallstone will release and concentrate all of the lipids present. An analysis of the lipids may lead to information about the cause of the gallstones.

27.49 Ubiquitin, a protein with 76 amino acids, is used to label damaged proteins that must be degraded. Ubiquitin is linked to targeted proteins by forming an amide bond between the carboxyl terminus of ubiquitin (Gly) to a side-chain amino group on a lysine residue of the doomed protein.

27.51 The reaction is a transamination:

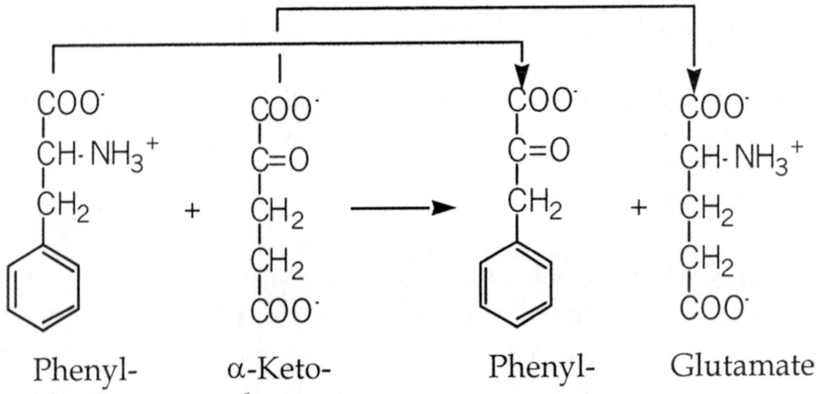

Phenyl-alanine + α-Keto-glutarate ⟶ Phenyl-pyruvate + Glutamate

27.53 The presence of a high concentration of ketone bodies in the urine of a patient is usually indicative of diabetes. However, before the disease can be confirmed, other, more detailed tests must be completed as fasting or starvation or special dieting can also increase ketone bodies.

27.55 (a) NAD^+ participates in step # 5, the oxidation of glyceraldehyde 3-phosphate to 1,3-bisphosphoglycerate and step # 12, the oxidation of pyruvate to acetyl CoA.
(b) NADH participates in steps # 10 and # 11, reduction of pyruvate to lactate and ethanol.
(c) If one considers the path from glucose to lactate, then there is no net gain of the cofactors. If one considers the path from glucose to pyruvate, then there is a gain of two NADH per glucose. Pyruvate to acetyl CoA would add another two NADH per glucose.

27.57 The amino acids Ala, Gly, and Ser are glucogenic; that is, their carbon atoms may be used to synthesize glucose, thus relieving hypoglycemia (low blood sugar).

27.59 Products of the transamination reaction of Ala and oxaloacetate:

$$
\begin{array}{ccccc}
COO^- & & COO^- & & COO^- & & COO^- \\
CH\cdot NH_3^+ & + & C{=}O & \longrightarrow & C{=}O & + & CH\cdot NH_3^+ \\
CH_3 & & CH_2 & & CH_3 & & CH_2 \\
& & COO^- & & & & COO^- \\
\text{Alanine} & & \text{Oxalo-} & & \text{Pyruvate} & & \text{Aspartate} \\
& & \text{acetate} & & & &
\end{array}
$$

27.61 A number of metabolic processes could occur with the radioactive fatty acid so different molecules should be analyzed. Some of the 'hot' fatty acid could be stored in triglycerides in fat tissue; some radioactivity would be in acetyl CoA after β oxidation of the fatty acids; and some would be in carbon dioxide (released from citric acid cycle).

27.63 The urea cycle is an energy-consuming pathway as it requires 3 molecules of ATP (four phosphate anhydride bonds) to produce a single urea from carbon dioxide and two ammonium ions.

27.65 (a) The β oxidation of lauric acid (12 carbons) requires 5 turns.
(b) Palmitic acid (16 carbons) requires 7 turns.

Chapter 28 Biosynthetic Pathways

28.1 Your text states in Section 28.1 several reasons why anabolic pathways are different from catabolic: (1) Duplication of pathways adds flexibility. If the normal biosynthetic pathway is blocked, the body can use the reverse of the catabolic pathway to make the necessary metabolites. (2) Separate pathways allow the body to overcome the control of reactions by reactant concentration (Le Chatelier's principle). (3) Different pathways provide for separate regulation of each pathway. Although there are many differences between anabolism and catabolism, we will also note similarities that allow for coordinated regulation and proper balancing of concentrations.

28.3 The cellular concentration of inorganic phosphate, a reagent used in phosphorylation reactions, is very high; therefore, the reaction is driven in the direction of glycogen breakdown. In order to ensure the presence of glycogen when needed, it must have an alternate synthetic pathway.

28.5 The major difference between the overall reactions of photosynthesis and respiration is the direction of the reactions. They are the reverse of each other:

$$6CO_2 + 6H_2O \rightarrow C_6H_{12}O_6 + 6O_2 \quad \text{photosynthesis}$$
$$C_6H_{12}O_6 + 6O_2 \rightarrow 6CO_2 + 6H_2O \quad \text{respiration}$$

28.7 A compound that can be used for gluconeogenesis:
   (a) From glycolysis: pyruvate
   (b) From the citric acid cycle: oxaloacetate
   (c) From amino acid oxidation: alanine

28.9 The brain obtains most of its energy from glucose that is supplied by the blood. The brain has little or no capacity to store glucose in glycogen. During starvation, glucose for the brain will come from glycogen that is stored primarily in the liver. Because glucose concentrations are very low in starvation, the liver glycogen is synthesized from glucose that is produced from pyruvate, lactate, amino acids, etc. The glucose is formed by gluconeogenesis. Brain cells are also able to obtain some energy from degradation of the ketone bodies.

28.11 Maltose is a disaccharide that is composed of two glucose units linked by an α-1,4-glycosidic bond (Section 19.5C). The enzyme is glycogen synthase.

$$\text{UDP-glucose} + \text{glucose} \longrightarrow \text{maltose} + \text{UDP}$$

28.13 Uridine triphosphate (UTP) is a nucleoside triphosphate similar to ATP. The constituents

Chapter 28 Biosynthetic Pathways

are: a nitrogen base, uracil; a sugar, ribose; and three phosphates.

28.15 (a) The biosynthesis of fatty acids occurs primarily in the cell cytoplasm. Here acetyl CoA is used to make palmitoyl CoA. Extension of the carbon chain to stearate and desaturation to form carbon-carbon double bonds occurs in mitochondria and the endoplasmic reticulum.
(b) Fatty acid catabolism does not occur in the same location as anabolism. The enzymes for β oxidation are located in the mitochondrial matrix.

28.17 In fatty acid synthesis, the compound that is added repeatedly to the enzyme, fatty acid synthase, is malonyl CoA, which has a three-carbon chain.

28.19 The carbon dioxide is released from malonyl CoA which leads to the addition of two carbons to the growing fatty acid chain.

28.21 If one considers only what is happening to the fatty acid, removal of two hydrogens and two electrons, then it looks like oxidation only. However, the reaction is much more complex and involves a cofactor, NADH and the substrate, oxygen. Both the fatty acid and NADH undergo two-electron reduction. The four electrons and protons are used to reduce oxygen to water:
$$O_2 + 4H^+ + 4e^- \longrightarrow 2H_2O$$

28.23 The only structural difference between NADH and NADPH is the presence of a phosphate group on one of the ribose units of NADPH. When considering the binding of NADPH to an NADH-requiring enzyme, two factors are important—size and charge differences. The phosphate adds a rather bulky group to the NADPH and the NADH binding site may not be able to accommodate the larger size of the cofactor. In terms of charge, NADPH has two negative charges not present in NADH. The NADH binding site may have amino acid residues that have negatively-charged side chains like Glu or Asp. These would repel NADPH, but could hydrogen bond to the free hydroxyl group in NADH.

28.25 Humans have enzymes that catalyze the oxidation of saturated fatty acids to mono-unsaturated fatty acids with the double bond between carbons 9 and 10. For example, we can make palmitoleic acid from palmitate and oleic acid from stearic acid. Humans do not have enzymes that introduce a double bond beyond the 10th carbon. Therefore humans cannot make linoleic (double bonds at carbons 9-10 and 12-13) or linolenic acid (double bonds at carbons 9-10, 12-13, and 15-16). Those fatty acids are essential in the diet.

28.27 To make a glucoceramide, sphingosine is reacted with an acyl CoA that adds a fatty acid in amide linkage. Glucose is added to the hydroxyl group of sphingosine using the activated form, UDP-glucose (Section 20.8).

28.29 All of the carbons in cholesterol orginate in acetyl CoA. An important intermediate in the synthesis of the steroid is a C-5 fragment called isopentenyl pyrophosphate:

$$3 \text{AcetylCoA} \longrightarrow \text{mevalonate} \longrightarrow \text{isopentenyl pyrophosphate} + CO_2$$

C-3   C-6   C-5

28.31 An amino acid is synthesized by the reverse of oxidative deamination (Section 28.5). The amino acid product is aspartate.

28.33 The products of the transamination reaction shown are valine and α-ketoglutarate.

$$(CH_3)_2 CH \cdot \overset{\overset{\displaystyle O}{\|}}{C} \cdot COO^- \ + \ ^-OOC \cdot CH_2 \cdot CH_2 \cdot \underset{\underset{\displaystyle NH_3{}^+}{|}}{CH} \cdot COO^- \longrightarrow$$

The keto form   Glutamate
of valine

$$(CH_3)_2 CH \cdot \underset{\underset{\displaystyle NH_3{}^+}{|}}{CH} \cdot COO^- \ + \ ^-OOC \cdot CH_2 \cdot CH_2 \cdot \overset{\overset{\displaystyle O}{\|}}{C} \cdot COO^-$$

Valine   α-Ketoglutarate

28.35 The carbon dioxide that is used to make carbohydrates in plants is reduced by the cofactor NADPH.

28.37 (a) The colored urine of blue diaper syndrome is caused by indigo blue dye.
(b) It is formed from the amino acid tryptophan.

28.39 The C-3 fragment, a malonyl group, is carried by ACP, acyl carrier protein.

28.41 No. Ras protein is a GTP binding protein involved in signal transduction only in its prenylated form. It is an oncogene, thus mutations may cause cancer. Some mutations that inhibit prenylation of Ras cause the cell to die. Other mutations of prenylated Ras may cause uncontrolled cell growth.

28.43 Towards the formation of α-hydroxybutyrate. When the human body is exposed to cold temperatures, energy metabolism must be increased to generate heat. Increasing the concentration of mainstream metabolites would do this. α-Ketoglutarate, resulting from the deamination of glutamate, would feed directly into the citric acid cycle and enhance metabolic rates.

Chapter 28 Biosynthetic Pathways

28.45 The C-10 and C-15 intermediates in cholesterol synthesis are geranyl pyrophosphate and farnesyl pyrophosphate, respectively.

28.47 The statement is also true for photosynthesis. The overall reaction for photosynthesis is shown in Problem 28.5 and Reaction 28.3. The carbon in carbon dioxide is in the most highly oxidized form. In the biosynthesis of carbohydrates, the carbon atoms become reduced to aldehyde and alcohol functional groups. Both of these groups come as a result of the reduction of the carbon dioxide.

28.49 Fatty acid synthase produces even-numbered chains up to 16 carbons, palmitate. Elongation of the C-16 acids occurs by a different process in mitochondria and the endoplasmic reticulum.

Chapter 29 Nutrition

29.1 No, nutrient requirements vary from person to person. The recommended daily allowances (RDA) stated by the government are average values and they assume a wide range of needs in different individuals. Some of the factors that change dietary needs include age, level of activity, genetics, and geographical region of residence.

29.3 Sodium benzoate is not catabolized by the body; therefore, it does not comply with the definition of a nutrient—components of food that provide growth, replacement, and energy. Calcium propionate enters mainstream metabolism by conversion to succinyl CoA and catabolism by the citric acid cycle.

29.5 The Nutrition Facts label found on all foods must list the percentage of daily values for four important nutrients: vitamins A and C, calcium, and iron.

29.7 Fiber is an important non-nutrient found in some foods. It is the indigestible portion of fruits, vegetables, and grains. Chemically, fiber is cellulose, a polysaccharide that cannot be degraded by humans. It is important for proper operation of dietary processes, especially in the colon.

29.9 The basal caloric requirement is calculated assuming the body is completely at rest. Because most of us perform some activity, we need more calories than this basic minimum. The caloric intake of 2100 for a young woman is a peak requirement. The difference between the basic level and the peak is needed to produce energy for activity.

29.11 Assume that each pound of body fat is equivalent to 3500 Cal. Therefore, the total number of calories that must be deleted from the diet is 3500 Cal/lb x 20 lb = 70,000 Cal. Because this must be done in 60 days, the amount restricted each day is 70,000 Cal ÷ 60 days = 1167 Cal/day. The caloric intake each day should be 3000 − 1167 = 1833 Cal.

29.13 Water makes up about 60 % of our body weight; therefore, one might assume she/he can lose weight by taking diuretics to enhance water release. However, this would only serve as a "quick fix" as the weight would return rapidly. In addition, it could be dangerous because the body needs a constant supply of water. The only way to lose weight is to reduce the level of body fat.

29.15 Amylose is a storage polysaccharide that is one of the components in starch. Chemically it is an unbranched polysaccharide composed of glucose residues linked by α-1,4-glycosidic bonds. α-Amylase is an enzyme that catalyzes the hydrolysis of the glycosidic bonds at random sites in amylose. Thus the product would be different-sized, polysaccharide fragments much smaller than the original amylose molecules.

29.17 No. Dietary maltose, the disaccharide composed of glucose units linked by an α-1,4-glyco-
sidic bond, is rapidly hydrolyzed in the stomach and small intestines. By the time it
reaches the blood, it is the monosaccharide glucose.

29.19 Arachidonic acid is produced from the essential, unsaturated fatty acid, linoleic acid.

29.21 No, lipases degrade neither cholesterol nor fatty acids. Lipases catalyze the hydrolysis
of the ester bonds in triglycerides, releasing free fatty acids and glycerol.

29.23 Yes, it is possible for a vegetarian to obtain a sufficient supply of adequate proteins;
however, the person must be very knowledgeable about the amino acid content of
vegetables. It is very difficult to find a single vegetable that has complete proteins, that is,
proteins with every essential amino acid. It is important for a vegetarian to eat a variety
of vegetables in a proper balance so that all of the essential amino acids, fats, and
nutrients are present.

29.25 Dietary proteins begin degradation in the stomach that contains HCl in a concentration
of about 0.5 %. Trypsin is a protease present in the small intestines that continues protein
digestion after the stomach. Stomach HCl denatures dietary protein and causes somewhat
random hydrolysis of the amide bonds in the protein. Fragments of the protein are
produced. Trypsin catalyzes hydrolysis of peptide bonds only on the carboxyl side of the
amino acids Arg and Lys.

29.27 The rice/water diet provides sufficient calories for the basal caloric requirement;
however, the diet is lacking in important nutrients such as some essential amino acids
(Thr, Lys), essential fats, vitamins, and minerals. It is expected that many of the prisoners
will develop deficiency diseases in the near future.

29.29 Limes provided sailors with a supply of vitamin C to prevent scurvy.

29.31 Vitamin K is essential for proper blood clotting.

29.33 The only disease that has been proven scientifically to be prevented by vitamin C is
scurvy.

29.35 Vitamins E and C, and the carotenoids may have significant effects on respiratory health.
This may be due to their activity as antioxidants.

29.37 There is a sulfur atom in biotin and in vitamin B-1 (also called thiamine).

29.39 Intravenous feeding with glucose may provide a patient with sufficient calories, but it is
dangerously deficient in vitamins, essential fats, and essential amino acids. In other words,

it is not a balanced diet.

29.41 Aspartame, the artificial sweetener, is the methyl ester of the dipeptide Asp-Phe. The methyl esters of Phe-Asp and Asp-Phe differ in the sequence of their amino acids. In aspartame, Asp is the N-terminus, whereas in the other dipeptide, Phe is the N-terminus.

29.43 Both olestra and sucralose have a backbone of sucrose, a disaccharide containing glucose and fructose (see Chemical Connections 29C).

29.45 Chlorine and ozone are used to disinfect public water supplies.

29.47 Megadoses (2 g/day) of nicotinamide, vitamin B, are used to treat Bullus pemphigoid.

29.49 The primary chemical process involved in digestion is hydrolysis of large food molecules (carbohydrates, fats, proteins) to their smaller precursors. Polysaccharides are cleaved by hydrolysis of glycosidic bonds to yield mono-, di-, and oligosaccharides which usually end up as free glucose. The ester bonds in triglycerides are hydrolyzed to form fatty acids and glycerol. The amide bonds in proteins are hydrolyzed to yield amino acids.

29.51 The polysaccharide amylose has no branches so the debranching enzyme does not act on it as a substrate.

29.53 Human insulin is a polypeptide hormone comprised of 51 amino acids. If it is administered orally, it is degraded by hydrolysis in the stomach and small intestines before it reaches its target cells.

29.55 A low-phenylalanine diet is required for individuals with the genetic disease PKU (see Chemical Connections 27F). These individuals lack the enzyme phenylalanine hydroxylase and are unable to metabolize the amino acid Phe properly. Because aspartame contains Phe that is released by digestion in the stomach, the sweetener must be restricted from the diet.

29.57 No. Although research shows that arsenic may be a micronutrient in some organisms, amounts above natural levels are very toxic in humans.

Chapter 30 Immunochemistry

30.1 Examples of external innate immunity include action by the skin, tears, and mucus.

30.3 The skin fights infection by providing a barrier against penetration of pathogens. The skin also secretes lactic acid and fatty acids, both of which create a low pH thus inhibiting bacterial growth.

30.5 Innate immunity processes have little ability to change in response to immune dangers. The key features of adaptive (acquired) immunity are specificity and memory. The acquired immune system uses antibody molecules designed for each type of invader. In a second encounter with the same danger, the response is more rapid and more prolonged than the first.

30.7 T cells originate in the bone marrow, but grow and develop in the thymus gland. B cells originate and grow in the bone marrow.

30.9 Macrophages are the first cells in the blood that encounter potential threats to the system. They attack virtually anything that is not recognized as part of the body including pathogens, cancer cells, and damaged tissue. Macrophages engulf an invading bacteria or virus and kill it using nitric oxide (NO, see Table 30.1).

30.11 T cells recognize peptide/protein antigens.

30.13 Class II MHC molecules pick up damaged antigens. A targeted antigen is first processed in lysosomes where it is degraded by proteolytic enzymes. An enzyme, GILT, reduces the disulfide bridges of the antigen. The reduced peptide antigens unfold and are further degraded by proteases. The peptide fragments remaining serve as epitopes that are recognized by class II MHC molecules (Figure 30.4).

30.15 MHC molecules are transmembrane proteins that belong to the immunoglobulin superfamily. They have peptide-binding variable domains. Class I molecules are single polypeptides, whereas, class II are protein dimers. They are originally present inside cells until antigens are detected, then they move to the surface membrane.

30.17 IgA molecules are found mainly in body secretions: tears, milk, and mucus. They serve as "front-line" immunoglobins as they attack invading agents before they get into the blood stream. IgE molecules respond to allergens that cause conditions like asthma and hay fever. IgG molecules are the most important antibodies in the blood where they recognize and strongly bind antigens.

Chapter 30 Immunochemistry

30.19 The blood carries antibodies against foreign substances. When antibodies bind targets, they cause aggregration because the antibodies have two binding sites on the two forks of the Y-shaped molecule. Transfusion using a blood of different type would activate the immune system and cause extreme aggregation of cells. Two blood cells may be linked by one antibody leading to an insoluble network of blood cells (Chemical Connections 19D):

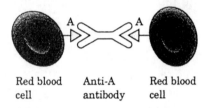

Red blood   Anti-A   Red blood
cell   antibody   cell

30.21 Each immunoglobulin molecule is composed of four protein chains, two identical light chains and two identical heavy chains. They are arranged into a Y-shaped tetramer (Figure 30.5). The subunits of the tetramer are held together by covalent disulfide bonds between Cys residues and also by noncovalent bonds including hydrogen bonding and hydrophobic interactions.

30.23 We start with the assumption that the two different MCAs are directed against a common antigen. Thus, they would have a different specificity, but structurally they would have identical constant regions and different variable regions (antigen binding sites).

30.25 During B cell development the variable regions of the H chains are assembled by V(J)D recombination. This creates a new V(J)D gene that provides great diversity because of the large number of combinations possible. When the body encounters a new antigen, mutations will be induced into the new genes (Section 30.4D).

30.27 T cells carry on their surfaces unique receptor proteins that are specific for antigens. These receptors (TcR), members of the immunoglobulin superfamily, have constant and variable regions. They are anchored in the T cell membrane by hydrophobic interactions. They are not able to bind antigens alone, but they need additional protein molecules called cluster determinants that act as coreceptors. When TcR molecules combine with cluster determinant proteins, they form T cell receptor complexes (TcR complex).

30.29 The components of the T cell receptor complexes are (1) accessory protein molecules called cluster determinants and (2) the T cell receptor.

30.31 The adhesion molecule in the TcR complex that assists HIV infection is the cluster determinant 4 (CD4).

Chapter 30 Immunochemistry

30.33 Cytokines are low molecular weight glycoproteins that are produced in one cell type and control the immune response in other cell types.

30.35 (a) TNF = tumor necrosis factor
(b) IL = interleukins
(c) EGF = epidermal growth factor

30.37 Chemokines are low molecular weight proteins that have distinct disulfide linkages. All chemokines have four Cys residues that form two disulfide bonds, Cys1-Cys3 and Cys2-Cys4.

30.39 Chemokines all have cysteine residues.

30.41 When a normal cell becomes malignant, the epitopes that indicated a healthy cell diminish and unusual epitopes begin to present themselves on the surface of the transformed cells. The effect is that fewer inhibitory receptors of macrophages bind to the surface of the cell (Section 30.8B).

30.43 When a normal healthy cell encounters a macrophage, the inhibitory receptors on the macrophage recognize the epitope of the normal cell, bind to it, and prevent the attack.

30.45 The glucocorticoids suppress the immune response to allergens that may cause discomfort or even death to some individuals. Glucocorticoids inhibit the synthesis of TcR by directly interacting with the gene or indirectly interacting with transcription factors.

30.47 MCA treatment for cancer is better than chemotherapy because MCAs are much more specific for the cancer cells. Treatment is directed toward just the cancerous cells, not all cells in general like chemotherapy. In addition, MCA therapy shows minimal toxicity.

30.49 Plasma cells are derived from B cells after the B cells have been exposed to an antigen. The primary role of plasma cells is to produce antibodies. Vaccination of an individual changes blood lymphocytes into plasma cells. Some lymphocytes are transformed into memory cells rather than plasma cells. Memory cells do not secrete antibodies. Rather, they store them for later use.

30.51 Chemokines play an important role in leukocyte movement. Binding of a chemokine to its receptor on the leukocyte causes changes inside the cell resulting in structural changes (they become flattened) and they are allowed to migrate. They can squeeze through the gaps between endothelial cells and out of the blood vessel to the site of injury.

30.53 The B cells of the afflicted person, which do not recognize self, make the high level of antibody.

30.55 Lymphatic capillary vessels enter the lymphoid organs—the thymus, spleen, the thoracic duct, and the lymph nodes.

30.57 Chemokines (or more generally cytokines) help leukocytes migrate out of a blood vessel to the site of injury. Cytokines help the proliferation of leukocytes.

30.59 A compound called 12:13 dEpoB, a derivative of epothilon B, is being studied as an anticancer vaccine.

30.61 Tumor necrosis factor receptors are located on the surfaces of several cell types, but especially on tumor cells.

Chapter 31 Body Fluids

<u>31.1</u> The blood-brain barrier refers to the limited exchange of molecules between blood and intersitial fluid of the brain. Molecules that are able to diffuse through the barrier include water, oxygen, glucose, small alcohols, and most anesthetics. The barrier is only slightly permeable to charged ions like sodium, potassium, and chloride.

<u>31.3</u> (a) As the blood plasma circulates through the body, it is in constant contact with other body fluids.
(b) The contact is through the semipermeable membranes of the blood vessels (see Figure 30.2). The membrane allows blood to exchange biochemicals with other fluids like lymph and interstitial fluids.

<u>31.5</u> Erythrocytes are the red blood cells. Their main function is to transport oxygen from the lungs to living cells throughout the body and to carry a waste product of cells, carbon dioxide, to the lungs.

<u>31.7</u> (a) Interstitial fluid is the aqueous-based liquid that surrounds most cells and fills the spaces between them.
(b) Blood plasma is the noncellular portion of blood that is the supernatant after centrifugation of whole blood. Cellular elements are in the sediment after centrifugation. Plasma is water-based and flows in the arteries and veins transporting nutrients to cells throughout the body.
(c) Serum is the fluid remaining after blood has been allowed to clot. Serum contains all the same components as plasma except fibrinogen is present in plasma, but not in serum. The fibrinogen has aggregated (polymerized) in the clot.
(d) The cellular elements of blood include erythrocytes, leukocytes, and platelets.

<u>31.9</u> When whole blood is centrifuged, the cellular elements sediment to the bottom of the tube. The remaining supernatant is the plasma.

<u>31.11</u> The immunoglobulins, principally IgG antibodies, protect against infection.

<u>31.13</u> A hemoglobin molecule is composed of four subunits, two chains are called α and two chains are identified as β. Each of the protein subunits has a heme prosthetic group which binds an oxygen molecule to the ferrous ion. Therefore, each molecule of hemoglobin can carry a full capacity of four molecules of oxygen.

<u>31.15</u> Ferrous ions, Fe^{2+}, the metal ions in each of the four heme groups in hemoglobin, are used to define oxygen saturation. When hemoglobin is fully saturated (100%), each of the four ferrous ions has an oxygen molecule bound to it.

31.17 See Section 21.9B for a review of the quaternary structure of hemoglobin. The four subunits are held together in a specific quarternary structure by noncovalent bonding including electrostatic interactions, hydrogen bonding, and hydrophobic interactions. These interactions bring all four subunits in contact with each other. When the first oxygen binds to a heme site, that subunit undergoes a change in three dimensional shape. This movement induces the same shape change in an adjoining subunit, thus making it easier for the second oxygen molecule to bind to the site. The oxygen affinity of the second subunit is enhanced. This increased affinity continues with each step so each oxygen is added with greater ease. We call this enhanced binding a cooperative effect.

31.19 (a) Carbon dioxide is bound to the α-amino group (N-terminus) of each of the four subunits of hemoglobin. (b) The complex produced is called carbaminohemoglobin: $Hb-NH-COO^-$.

31.21 The standard equilibrium constant for the reaction is defined as the concentration of the carbonic acid product divided by the concentrations of the reactants (water and carbon dioxide) (see Section 7.6). If we assume that the factor for water concentration is included in the equilibrium constant, than the equilibrium constant is equal to [carbonic acid] ÷ [carbon dioxide] = 70 ÷ 5 = 14.

31.23 Water is reabsorbed in the kidney in the region called the proximal tubule, a part of a nephron.

31.25 The normal constituents of urine, in addition to the nitrogen-containing waste products, are water and inorganic ions—Na^+, Ca^{2+}, Mg^{2+}, Cl^-, PO_4^{3-}, SO_4^{2-}, and HCO_3^-.

31.27 Protons in the blood are neutralized by bicarbonate ions that are part of the carbonic acid buffer system (Section 8.11D).

31.29 In the presence of the hormone vasopressin, water is reabsorbed in the distal tubules and the collecting tubules of the kidney.

31.31 Kidney function involves balancing water between filtration (for release by urine) and reabsorption. Diuretics are drugs that shift the balance toward filtration, thus they act to increase urine volume. More water is released in the urine, but the amount of dissolved solutes remains the same so urine is, in effect, diluted. Specific gravity is defined as the density of a substance compared to water as a standard. The concentration of the solutes in urine diluted by water is lower than normal, thus the specific gravity is lower.

31.33 In order for nutrients like glucose to flow from blood into the interstitial fluids and into cells, blood pressure must be greater than the osmotic pressure at capillary ends (18 mm Hg). A pressure of 12 mm Hg would not allow a flow of glucose from the capillary to the

Chapter 31 Body Fluids

interstitial fluids.

31.35 ACE inhibitors reduce the production of angiotensin, a hormone that raises blood pressure by vasoconstriction. Diuretics enhance urine excretion so they decrease blood volume and thus lower blood pressure.

31.37 Blood platelets are anchored first on the collagen of the injured site.

31.39 When a blood vessel is injured, the first agents to the scene are the platelets. They are formed in the bone marrow and constantly circulate in the blood.

31.41 A symptom of enlarged prostate gland is restricted flow of urine caused by constriction of the uretor.

31.43 Blood pressure may be lowered by drugs that block calcium channels. For muscle contraction to occur, calcium ions must pass through muscle cell membranes. Calcium ions are transported through special channels formed by intramembrane proteins. If these channels are blocked by the drugs, the heart muscles will contract less frequently, thus pumping less blood and lowering pressure.

31.45 Serum is lacking the plasma protein fibrinogen, thus the total protein content of serum is less than that of plasma. The actual amount of albumin, by weight, is the same in both, but the concentration of albumin is measured by mg albumin ÷ mg total protein. The numerator remains the same for serum and plasma, but the denominator is less for serum. Therefore, albumin concentration is higher in serum than in plasma.

31.47 Acidosis, an increase of protons in the blood (lowering of pH), changes hemoglobin's affinity for oxygen (the Bohr effect). The oxygen-carrying ability of hemoglobin under acidic conditions will be less than under normal physiological conditions, because protons enhance the release of oxygen.

31.49 It is expected that the protein hormone insulin, with 51 amino acids, some with charged side chains, and a molecular weight greater than 5000, would not pass through the blood-brain barrier.

31.51 Water retention is regulated by the hormone vasopressin, a small peptide. Vasopressin acts to reabsorb water by the proximal tubule, the distal tubule, and the collecting tubule.

31.53 The oxygen dissociation curve (Figure 31.3) for fetal hemoglobin (HbF) is higher and shifted to the left when compared to normal adult hemoglobin (HbA). This means that lower pressures of oxygen are required to saturate HbF. We conclude that HbF has a higher affinity for oxygen than does HbA. Therefore, oxygen will prefer to bind to HbF. HbF has a slightly different primary structure that increases its oxygen affinity.